SpringerBriefs in Microbiology

For further volumes:
http://www.springer.com/series/8911

Kumar Sudesh

Polyhydroxyalkanoates from Palm Oil: Biodegradable Plastics

 Springer

Dr. Kumar Sudesh
School of Biological Sciences
Universiti Sains Malaysia
Penang
Malaysia

ISSN 2191-5385   ISSN 2191-5393   (electronic)
ISBN 978-3-642-33538-9   ISBN 978-3-642-33539-6   (eBook)
DOI 10.1007/978-3-642-33539-6
Springer Heidelberg New York Dordrecht London

Library of Congress Control Number: 2012949126

Printed on acid-free paper

Springer is part of Springer Science+Business Media (www.springer.com)

Preface

Polyhydroxyalkanoates (PHAs) are very interesting polyesters synthesized by many types of bacteria. Numerous researchers from all over the world have carried out various studies on PHAs. There is already a wealth of knowledge about all aspects of PHAs in the literatures.

In this book, the focus is on the relatively recent efforts to use vegetable oils, especially palm oil and its by-products to synthesize PHAs. Palm oil is the world's most efficiently produced vegetable oil and Malaysia has been the pioneer in developing palm oil as a sustainable source of edible oil. Because of the high productivity of palm oil it costs less than other commercial vegetable oils. Therefore, it has been the preferred oil by most people from the low-income group. The production of palm oil is expected to increase to fulfill the growing demand. Besides Malaysia, Indonesia is also now a major producer of palm oil.

The palm oil industry generates large quantities of by-products and wastes rich in fatty acids that can be developed into potential feedstock for biotechnological applications such as for the production of PHA by microbial fermentation. Studies have also shown that the yields of PHAs from vegetable oils are generally better than those from sugars or other feedstock. Besides the yield of PHAs, there are many other factors that one will have to consider for large-scale production of PHAs. Of particular importance is the sustainability of the entire process of converting palm oil-based feedstock to PHA.

Malaysia is committed to the production of palm oil in a sustainable manner. The majority of Malaysians are also environmentally conscious and know the importance of biodiversity and forest conservation. Therefore, the pros and cons of developing palm oil-based feedstock for PHA production are being carefully scrutinized. This book is an attempt to provide a holistic view of the challenges involved in using palm oil and its by-products for the production of PHAs. In addition, several new applications for PHAs are also described.

This book was prepared in the midst of many other equally demanding tasks and therefore the help of many of my laboratory members was crucial. I am especially grateful to Dr. Sridewi Nanthini for compiling all the information necessary for this book. Mr. Yoga S. Salim had painstakingly drawn all the chemical structures of PHA monomers and Ms. Rathi Devi Nair did all the corrections based on inputs from all my graduate students. I am very grateful to all of them for their help in preparing this book.

Kumar Sudesh

Contents

Chapter 1
Introduction

Abstract Polyhydroxyalkanoate (PHA) is a plastic-like material synthesized by many bacteria. PHA serves as an energy and carbon storage compound for the bacteria. PHA can be extracted and purified from the bacterial cells and the resulting product resembles some commodity plastics such as polypropylene. Because PHA is a microbial product, there are natural enzymes that can degrade and decompose PHA. Therefore, PHA is an attractive material that can be developed as a bio-based and biodegradable plastic. In addition, PHA is also known to be biocompatible and can be used in medical devices and also as bioresorbable tissue engineering scaffolds. In this chapter, a brief introduction about PHA and the fermentation feedstock for its production are given.

Keywords Bio-based • Biodegradable • Microorganism • Palm oil • PHA Plastics • Polyhydroxyalkanoate • Polymer

Petroleum-derived plastics have contributed significantly to our modern lifestyle due to their favorable stability, durability, and suitable mechanical and thermal properties. Plastic materials have now become an integral part of our daily life. In fact, it is almost impossible to lead a normal life without plastics. From the clothes that we wear to the cars that we drive there are numerous components that are made of plastics. The food that we eat comes in plastic packages or containers. We pay for the food and almost everything else using plastic credit cards. When we fall sick and visit the doctor, we are prescribed medicines that come in plastic bottles or plastic zip lock bags. Many medical devices are made of plastics. Almost all electronic devices contain plastic components. Our children play with plastic toys and spend the first few years of their life with plastic diapers. We might also need similar diapers during the last few years of our life. Our dependence on plastics shows that plastics are preferred because of their safety, versatility, durability, and affordability. However, synthetic plastics are difficult to be disposed due to the lack of natural enzymes and biological processes that can efficiently degrade synthetic plastics. Incineration of plastics releases hazardous gases into the environment. Harmful chemical such as hydrogen cyanide can be formed from acrylonitrile-based plastics during the combustion of these plastics (Johnstone 1990;

K. Sudesh, *Polyhydroxyalkanoates from Palm Oil: Biodegradable Plastics*, SpringerBriefs in Microbiology, DOI: 10.1007/978-3-642-33539-6_1, © The Author(s) 2013

Atlas 1993). On the other hand, recycling might be a better alternative but it is a labor-intensive process. Categorization of a wide variety of plastics is a time-consuming process. This process becomes worse with the presence of additives such as pigment, coating, and fillers (Fletcher 1993).

The ever-growing need to curb plastic waste management problem has resulted in the search for an alternative to petroleum-based synthetic plastics. In addition, the fact that petroleum is a non-renewable resource, which would be depleted sooner or later, has also motivated the search for bio-based plastics from renewable resources. Bio-based plastics that are also biodegradable may offer a solution to the plastic waste management problem. Therefore, biodegradable polymers are investigated to replace common plastics (Song et al. 1999). Biodegradable plastics can be classified into three categories which are; chemically synthesized polymer, starch-based biodegradable plastics, and polyhydroxyalkanoates (PHAs) (Khanna and Srivastava 2005a). PHA is a microbial storage polyester, synthesized naturally by many types of bacteria. PHA is being considered as a potential renewable alternative to some petrochemical plastics. This is because the properties of PHA resemble the properties of some commercially available plastics (Sudesh et al. 2000). In addition, PHA is completely biodegradable in nature. The bio-based and biodegradable nature of PHA would have the long-term benefits of reducing plastic waste accumulation, global warming, pollution, and dependence on fossil fuels. The availability of cheap and renewable carbon feedstock, preferably bio-based, for efficient conversion into PHA would make the PHA products' prices competitive with their petroleum counterparts. For this purpose, plant oils have been investigated and were found to be very attractive carbon sources for large-scale PHA production. Plant oils yield higher PHA content in comparison with other tested substrates such as sugars, because of their complex mixture of triglycerides (Akiyama et al. 2003). Among the various plant oils, palm oil is the world's most efficiently produced oil. Malaysia and Indonesia are both major producers and exporters of palm oil in the world. The versatility of palm oil suggests its usage as edible oils as well as for the production of oleochemicals. The palm oil industry generates large quantities of by-products composed of triglycerides and fatty acids which are suitable for microbial utilization. As is the case for almost all new technological innovations, there are pros and cons in using plant oils for the commercial production of PHA. Numerous concerns have been raised about the merits of diverting food grade oil for PHA production at the expense of food supply on a global scale. In addition, an increase in the demand for plant oils may result in further expansion of oil palm plantations into forests and subsequently threatening wild life habitats and destroying precious biodiversities. This book reviews the use of palm oil and its by-products as renewable feedstock and to provide a future outlook on the sustainability of palm oil for PHA production. The production of PHA from other plant oils is also described. Finally, discussions on the production and characteristics of the various types of PHA produced from palm oil products and some new applications of the resulting polymers are also included in this book.

Chapter 2
Bio-Based and Biodegradable Polymers

Abstract Many types of biodegradable plastics are being developed in response
to the concerns over the accumulation of commodity plastics in the environment.
Polyhydroxyalkanoate (PHA) stands out as an attractive material because it can
be produced from renewable feedstock and subsequently it can be degraded
completely by microorganisms. Approximately, 150 structurally different monomers can be polymerized into PHAs by bacteria, giving rise to polymers with
diverse properties. Careful manipulation of the bacterial culture conditions and the
carbon feedstock allows the design and synthesis of tailor-made polymers for various applications. This chapter focuses on the genetics and biochemistry of PHA
biosynthesis in bacteria and recombinant organisms.

Keywords Degradation • Metabolic pathways • PHA synthase • PHA
granule • Renewable feedstock • Sugars • Transgenic plants • Triglycerides

Bio-based polymers are defined as polymers produced from biological renewable
resources and polymerized by chemical and/or biological methods. Bio-based polymers can be categorized into three groups: bio-chemosynthetic polymers [e.g.,
poly(lactic acid) (PLA), poly(butylene succinate), polyvinyl alcohol and polyg-
lycolic acid], biosynthetic polymers (bioplastics or naturally occurring polymers
[e.g., Polyhydroxyalkanoate (PHAs)]), and modified natural polymers (e.g., starch
polymers and cellulose derivatives) (Sudesh and Iwata 2008) (Table 2.1). For bio-
chemosynthetic polymers, the monomers are synthesized biologically and polymerized chemically. An example of bio chemosynthetic polymer is poly(lactic acid).
The lactic acid monomers which are produced by microbial fermentation are then
chemically polymerized into PLA (a process catalyzed by metal catalyst) (Jem et al.
2010). In contrast to PLA, PHA is completely produced by biological process. The
entire process of PHA biosynthesis right from the production of monomers and the
subsequent polymerization processes occurs in the bacterial cells. Various types of
bacterial enzymes act as the biological catalyst in the biosynthesis of PHA. Unlike
bio-chemosynthetic polymers and biosynthetic polymers, the natural polymers such
as starch need to be modified chemically and/or physically to enhance the polymer
structure and improve the thermal and mechanical properties (Hoover et al. 2010).

K. Sudesh, *Polyhydroxyalkanoates from Palm Oil: Biodegradable Plastics*,
SpringerBriefs in Microbiology, DOI: 10.1007/978-3-642-33539-6_2,
© The Author(s) 2013

3

Table 2.1 Some examples of bio-based and biodegradable polymers (Flieger et al. 2003; Nair and Laurencin 2007; Sudesh and Iwata 2008)

Category	Processes involved	Example	Biodegradability
Bio-chemosynthetic polymers	Biological synthesis of monomers and chemical polymerization	Poly(lactic acid) Poly(butylene succinate) Polyvinyl alcohol Polyglycolic acid Polythioesters	Hydrolytically degradable except crystalline poly(lactic acid) and polythioesters
Biosynthetic polymers	Biosynthesis of polymer by microorganisms	Poly(3-hydroxybutyrate)	Enzymatically and/or hydrolytically degradable
Modified natural polymers	Chemical modification of natural polymer	Starch polymer Cellulose derivatives Proteins	Enzymatically degradable

However, not all bio-based polymers are biodegradable. Some of the bio-based polymers such as crystalline PLA, cellulose derivatives, and polythioesters are not biodegradable (Steinbüchel 2005; Sudesh and Iwata 2008). Biodegradable polymers can be degraded hydrolytically and/or enzymatically, which involves the breakage of bonds that hold the monomers together in the polymer (Nair and Laurencin 2007). The natural polymers such as starch and proteins can be enzymatically degraded by various enzymes such as amylases and proteases, while the polymers possessing functional groups such as esters, anhydrides, carbonates, amides, and urea can be hydrolyzed (Flieger et al. 2003; Nair and Laurencin 2007).

Most naturally occurring polymers such as PHA undergo hydrolysis of ester bonds, which are catalyzed by intracellular or extracellular PHA depolymerase enzymes (Guérin et al. 2010). The extracellular depolymerase enzymes are secreted by various bacteria and fungi in order to break down the PHA into low molecular weight products that can be assimilated and metabolized by the microorganisms. The final products of PHA degradation under aerobic conditions are carbon dioxide and water while methane is produced under anaerobic conditions. Thin solvent-cast PHA films can be biodegraded completely on soil surface in less than 2 months under tropical conditions (Sudesh and Iwata 2008). If the films are buried in the soil, the degradation rate will be much faster because of greater exposure to soil microorganisms. Therefore, PHAs are good candidates for biodegradable plastics applications due to their similar properties to those of conventional plastics but with high biodegradability. The common homopolymer, poly(3-hydroxybutyrate) [P(3HB)], can be hydrolytically degraded to a normal blood constituent in human body; thus, it has a potential application as biomaterial (Nair and Laurencin 2007). However, because of the high crystallinity and the absence of PHA depolymerase in humans, this polymer degrades at a very slow rate. Another type of PHA, poly (3-hydroxybutyrate-*co*-3-hydroxyvalerate) [P(3HB-*co*-3HV)] with lower crystallinity than P(3HB) and higher rate of degradation is suggested as a better temporary substrate for tissue engineering such as bone tissue and epithelial tissue (Chen and Wu 2005; Köse et al. 2003). The PHA containing 4-hydroxybutyrate (4HB) monomer is probably the best to be used as biomaterials for tissue engineering because it can be hydrolyzed by the lipases of eukaryotes (Sudesh 2004).

2.1 Overview of Polyhydroxyalkanoate

PHAs are biopolyesters of various hydroxyalkanoates (HAs) synthesized as intracellular carbon and energy storage compounds by numerous bacteria (Lee and Choi 1999; Tang et al. 2008; Sudesh et al. 2011). Gram-positive and Gram-negative bacteria from at least 75 different genera are known to synthesize PHA (Reddy et al. 2003). These intracellular reserve polymers can be accumulated by some bacteria up to a maximum of 90 wt% of the cell dry weight (Madison and Huisman 1999a) under laboratory conditions such as during nutrient stress, which includes limitation of nitrogen, phosphorus, magnesium, or oxygen but in the presence of excess carbon supply (Anderson and Dawes 1990; Doi 1990; Kato et al. 1996; Madison and Huisman 1999). Under conditions that are not favorable for growth, the bacterial cells will assimilate the excess carbon sources for storage purposes. The assimilated carbon sources are then biochemically processed into 3-hydroxyalkanoic acid (3HA) monomer units and polymerized into high molecular weight water-insoluble PHA molecules that are stored in the form of inclusions or granules in the microbial cell cytoplasm.

The high refractivity of PHA granules allows it to be observed under phase-contrast light microscope as discrete inclusions of 0.2–0.5 μm in diameter (Sudesh et al. 2000; Sathesh and Murugesan 2010). The PHA molecules synthesized by bacteria have adequately high molecular weights to match the polymer qualities of conventional plastics such as polyethylene (PE) and polypropylene (PP) (Qin et al. 2007; Madison and Huisman 1999). This has drawn increasing attention on PHA as a substitute for petrochemical-based synthetic plastics as it can be thermally processed into various forms. PHAs are also advantageous over conventional plastics due to their biodegradability in natural environments (Sudesh and Iwata 2008). In addition, the use of renewable resources such as sucrose, starch, cellulose, triacylglycerols, palm oil, and activated sludge to supply microorganisms with various carbon substrates for the synthesis of PHA is also very attractive (Reddy et al. 2003; Ojumu et al. 2004).

Since the identification of P(3HB) in *Bacillus megaterium* by (Lemoigne 1926), 3-hydroxybutyrate (3HB) was regarded as the sole PHA monomeric unit. The presence of other monomer units such as 3-hydroxyvalerate (3HV) and 3-hydroxyhexanoate (3HHx) was only discovered almost 50 years later by (Wallen and Rohwedder 1974). To date, more than 150 different monomer constituents of PHA have been identified (Steinbüchel and Lütke-Eversloh 2003). Bacterial PHA can be divided into three main types depending on the number of carbon atoms in the monomeric units: short-chain-length (scl), medium-chain-length (mcl) and

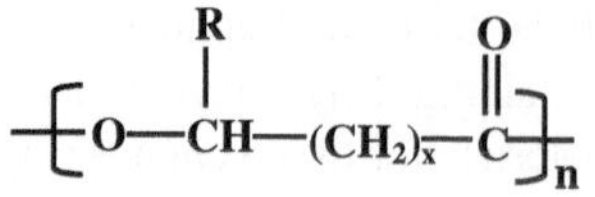

$$\left[O - \underset{\underset{\text{CH}}{|}}{\overset{\overset{R}{|}}{\text{CH}}} - (CH_2)_{\overline{x}} - \overset{\overset{O}{\parallel}}{C} \right]_n$$

Number of repeating units, x	Alkyl group, R	Polymer type
1	Hydrogen	Poly(3-hydroxypropionate)
	Methyl	Poly(3-hydroxybutyrate)
	Ethyl	Poly(3-hydroxyvalerate)
	Propyl	Poly(3-hydroxyhexanoate)
	Pentyl	Poly(3-hydroxyoctanoate)
	Nonyl	Poly(3-hydroxydodecanoate)
2	Hydrogen	Poly(4-hydroxybutyrate)
3	Hydrogen	Poly(5-hydroxyvalerate)

Fig. 2.1 General structures of polyhydroxyalkanoates (Lee 1996a)

a combination of scl-mcl. The scl-PHAs consist of 3–5 carbon atoms, mcl-PHAs have 6–14 carbon atoms whereas the number of carbon atoms in scl-mcl-PHAs can range from 3 to 14 per monomer (Li et al. 2007). While the homopolymer P(3HB) is the most widely studied scl-PHA, its copolymers containing 3HV, 3HHx, or 4HB monomers can also be synthesized (Fig. 2.1). In nature, scl-PHAs containing mainly 3HB units or mcl-PHAs containing 3-hydroxyoctanoate (3HO) and 3-hydroxydecanoate (3HD) are produced as the predominant mono-mers by most of the microbes (Anderson and Dawes 1990; Steinbüchel and Füchtenbusch 1998). The copolymers can be a more flexible and tougher plastics compared to the relatively stiff and brittle P(3HB). The usually elastomeric and sticky mcl-PHAs can even be modified to make rubbers (Suriyamongkol et al. 2007). Figure 2.2 shows the structures of HA constituents isolated from various microorganisms.

2.2 PHA Biosynthesis

PHA synthase (PhaC) is the key enzyme responsible for the polymerization of 3HA monomers (Qin et al. 2007; Pantazaki et al. 2009). Owing to the stere-ospecificity of this enzyme, all the 3HA monomer units are in the R configuration (Dawes and Senior 1973; Sudesh et al. 2000). PHA synthases are differentiated

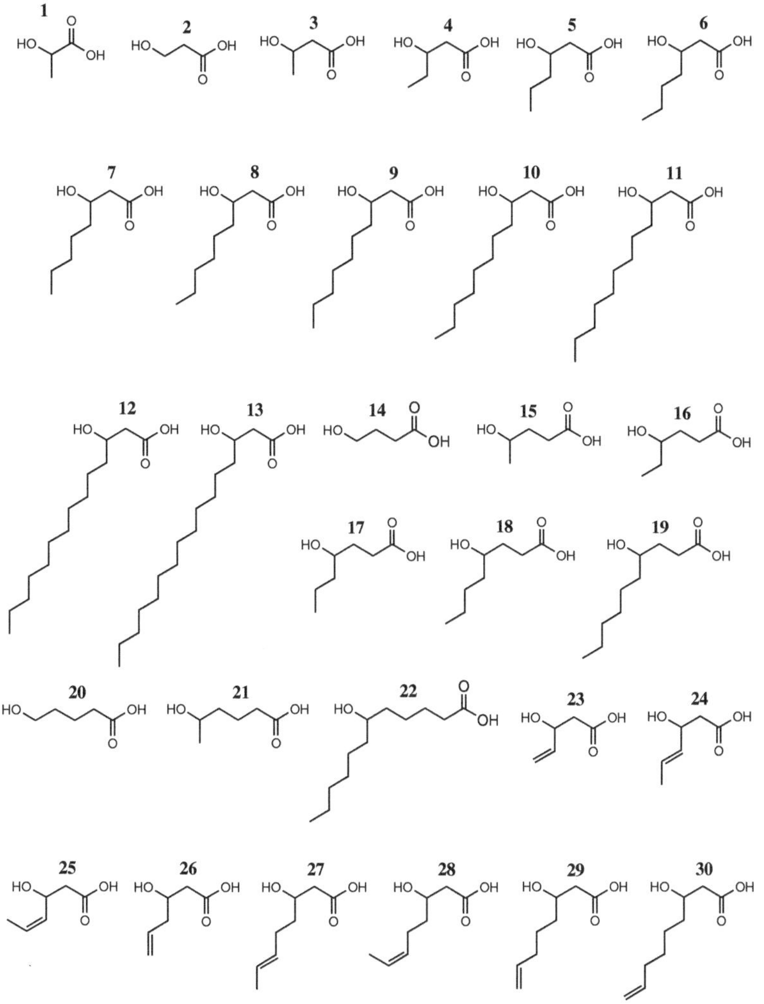

Fig. 2.3 Skeletal formula of HA constituents isolated from various microorganisms (adapted and updated from Steinbüchel and Valentin 1995). **#1**:Lactic acid (Valentin and Steinbüchel 1993), **#2**:3-hydroxypropionic acid (Doi et al. 1990), **#3**:3-hydroxybutyric acid (Lemoigne 1926), **#4**:3-hydroxyvaleric acid (Holmes et al. 1982), **#5**:3-hydroxyhexanoic acid (Lageveen et al. 1988), **#6**:3-hydroxyheptanoic acid (Findlay and White 1983; Lageveen et al. 1988), **#7**:3-hydroxyoctanoic acid (De Smet et al. 1983; Findlay and White 1983), **#8**:3-hydroxynonanoic acid (Lageveen et al. 1988), **#9**:3-hydroxydecanoic acid (Lageveen et al. 1988), **#10**:3-hydroxyundecanoic acid (Lageveen et al. 1988), **#11**:3-hydroxydodecanoic acid (Lageveen et al. 1988), **#12**:3-hydroxytetradecanoic acid (Lee et al. 1995), **#13**:3-hydroxyhexadecanoic acid (Lee et al. 1995), **#14**:4-hydroxybutyric acid (Kunioka et al. 1988), **#15**:4-hydroxyvaleric acid (Valentin et al. 1992), **#16**:4-hydroxyhexanoic acid (Valentin et al. 1994), **#17**:4-hydroxyheptanoic acid (Valentin et al. 1996), **#18**:4-hydroxyoctanoic acid (Valentin et al. 1996), **#19**:4-hydroxydecanoic acid (Eggink et al. 1995), **#20**:5-hydroxyvaleric acid (Doi et al. 1987b), **#21**:5-hydroxyhexanoic acid (Valentin et al. 1996), **#22**:6-hydroxydodecanoic acid (Eggink et al. 1995), **#23**:3-hydroxy-4-pentenoic acid (Lenz et al. 1992), **#24**:3-hydroxy-4-*trans*-hexenoic acid (Fritzsche et al. 1990b), **#25**:3-hydroxy-4-*cis*-hexenoic acid (Fritzsche et al. 1990b), **#26**:3-hydroxy-5-hexenoic acid (Fritzsche et al. 1990b), **#27**:3-hydroxy-6-*trans*-octenoic acid (Fritzsche et al. 1990b), **#28**:3-hydroxy-6-*cis*-octenoic acid (Fritzsche et al. 1990b), **#29**:3-hydroxy-7-octenoic acid (Fritzsche et al. 1990b), **#30**:3-hydroxy-8-nonenoic acid (Lageveen et al. 1988).

(continued)

#31:3-hydroxy-8-decenoic acid (Lageveen et al. 1988), #32:8-cyano-3-hydroxyoctanoic acid (Kim et al. 1998), #33:10-cyano-3-hydroxyundecanoic acid (Kim et al. 1998), #34:7-cyano-3-hydroxyheptanoic acid (Eggink et al. 1995), #35:9-cyano-3-hydroxynonanoic acid (Eggink et al. 1995), #36:3-hydroxy-6-*cis*-dodecenoic acid (Eggink et al. 1995), #37:3-hydroxy-5-*cis*-dodecenoic acid (Eggink et al.), #38:3-hydroxy-5-tetradecenoic acid (Eggink et al. 1995), #39:3-hydroxy-7-tetradecenoic acid (Choi and Yoon 1994), #40:3-hydroxy-5,8-*cis–cis*-tetradecenoic acid (Eggink et al. 1995), #41:3-hydroxy-4-methylvaleric acid (Findlay and White 1983), #42:3-hydroxy-4-methylhex-anoic acid (Fritzsche et al. 1990c), #43:3-hydroxy-5-methylhexanoic acid (Fritzsche et al. 1990c), #44:3-hydroxy-6-methylheptanoic acid (Findlay and White 1983), #45:3-hydroxy-4-methyloctanoic acid (Fritzsche et al. 1990c), #46:3-hydroxy-5-methyloctanoic acid (Fritzsche et al. 1990c), #47:3-hydroxy-6-methyloctanoic acid (Fritzsche et al. 1990c), #48:3-hydroxy-7-methyloctanoic acid (Fritzsche et al. 1990c), #49:3-hydroxy-6-methylnonanoic acid (Hazer et al. 1993), #50:3-hydroxy-7-methylnonanoic acid (Hazer et al. 1993), #51:3-hydroxy-8-methylnonanoic acid (Hazer et al. 1993), #52:3-hydroxy-7-methyldecanoic acid (Hazer et al. 1993), #53:3-hydroxy-9-methyldecanoic acid (Hazer et al. 1993), #54:3-hydroxy-7-methyl-6-octenoic acid (Choi and Yoon 1994), #55:Malic acid (Fischer et al. 1989), #56:3-hydroxysuccinic acid-methyl ester*, #57:3-hydroxyadipinic acid-methyl ester*, #58:3-hydroxysuberic acid-methyl ester (Lenz et al. 1992), #59:3-hydroxyazelaic acid-methyl ester (Lenz et al. 1992), #60:3-hydroxysebacic acid-methyl ester (Lenz et al. 1992), #61:3-hydroxysuberic acid-ethyl ester (Lenz et al. 1992), #62:3-hydroxysebacic acid-ethyl ester (Lenz et al. 1992), #63:3-hydroxypimelic acid-propyl ester*.

(continued)

#64:3-hydroxysebacic acid-benzyl ester*, #65:3-hydroxy-5-oxohexanoic acid (Jung et al. 2000), #66:3-hydroxy-7-ox-ooctanoic acid (Jung et al. 2000), #67:3-hydroxy-8-acetoxyoctanoic acid (Lenz et al. 1992), #68:3-hydroxy-9-acetoxyno-nanoic acid (Lenz et al. 1992), #69:3-hydroxy-4-benzoylbutyric acid (Kim et al. 1998), #70:3-hydroxy-5-benzoylvaleric acid (Kim et al. 1998), #71:3-hydroxy-6-benzoylhexanoic acid (Kim et al. 1998), #72:3-hydroxy-7-benzoylheptanoic acid (Kim et al. 1998),# 73:3-hydroxy-8-benzoyloctanoic acid (Kim et al. 1998), #74:3-hydroxy-5-(4-fluorobenzoyl) valeric acid (Yano et al. 2009), #75:3-Hydroxy-5-(2-thienoyl)valeric acid (Yano et al. 2003), #76:3-Hydroxy-6-(2-thienoyl)hexanoic acid (Yano et al. 2003), #77:3-Hydroxy-5-(2-thienyl-sulfanyl)valeric acid (Kenmoku et al. 2004), #78:3-hydroxy-6-(2-thienyl-sulfanyl)hexanoic acid (Kenmoku et al. 2004), #79:3-hydroxy-4-phenoxybutyric acid (Kim et al. 1996b), #80:3-hydroxy-5-phenoxyvaleric acid (Kim et al. 1996b), #81:3-hydroxy-6-phenoxyhexanoic acid (Kim et al. 1996b), #82:3-hydroxy-7-phenoxyheptanoic acid (Kim et al. 1996b), #83:3-hydroxy-8-phenoxyoctanoic acid (Kim et al. 1996b), #84:para-Cyanophenoxy-3-hydroxybutyric acid (Gross), #85:para-Cyanophenoxy-3-hydroxy-valeric acid (Gross), #86:para-Cyanophenoxy-3-hydroxyhexanoic acid (Gross), #87:para-Nitrophenoxy-3-hydrox-yhexanoic acid (Gross), #88:3-hydroxy-5-(3-fluorophenoxy)valeric acid (Takagi et al. 2004), #89:3-hydroxy-7-(3-fluorophenoxy)heptanoic acid (Takagi et al. 2004), #90:3-hydroxy-5-(4-fluorophenoxy)valeric acid (Takagi et al. 2004), #91:3-hydroxy-7-(4-fluorophenoxy)heptanoic acid (Takagi et al. 2004). *Scholz C. Personal communication in 1994.

(continued)

#92:3-hydroxy-5-(2,4-difluorophenoxy)valeric acid (Takagi et al. 2004), #93:3-hydroxy-5-para-tolylvaleric acid (Curley et al. 1996), #94:3-hydroxy-4-para-methylphenoxybutyric acid (Kim et al. 2000), #95:3-hydroxy-6-para-methylphenoxyhexanoic acid (Kim et al. 2000), #96:3-hydroxy-11-(bis(2-hydroxyethyl)-amino)-10-hydroxyundecanoic acid (Sparks and Scholz 2008), #97:3,6-epoxy-7-nonene-1,9-dioic acid (He et al. 1998), #98:3-hydroxy-5-phenylvaleric acid (Fritzsche et al. 1990a), #99:3-hydroxy-6-phenylhexanoic acid (Garcia et al. 1999), #100:3-hydroxy-8-phenyloctanoic acid (Garcia et al. 1999), #101:3-hydroxy-10-phenyldecanoic acid (Garcia et al. 1999), #102:3-hydroxy-5-(4-nitrophenyl)valeric acid (Arostegui et al. 1999), #103:3-hydroxy-5-(2,4-dinitrophenyl)valeric acid (Arostegui et al. 1999), #104:3-hydroxy-6-(4-vinylphenyl)hexanoic acid (Suzuki et al. 2004), #105:3-hydroxy-8-(4-vinylphenyl)octanoic acid (Suzuki et al. 2004), #106:3-hydroxy-10-(4-vinylphenyl)undecanoic acid (Suzuki et al. 2004), #107:3-hydroxy-5-(4-vinylphenyl)valeric acid (Suzuki et al. 2004), #108:3-hydroxy-7-(4-vinylphenyl)heptanoic acid (Suzuki et al. 2004), #109:3-hydroxy-9-(4-vinylphenyl)nonanoic acid (Suzuki et al. 2004), #110:3-hydroxy-4-[4-(methylsulfanyl)phenoxy]butyric acid (Kenmoku et al. 2002), #111:3-hydroxy-5-[4-(methylsulfanyl)phenoxy]valeric acid (Kenmoku et al. 2002), #112:3-hydroxy-6-[4-(methylsulfanyl)phenoxy]hexanoic acid (Kenmoku et al. 2002), #113:3-hydroxy-5-thiophenoxyvaleric acid (Takagi et al. 1999), #114:3-hydroxy-6-thiophenoxyhexanoic acid (Takagi et al. 1999).

(continued)

#**115**:3-hydroxy-7-thiophenoxyheptanoic acid (Takagi et al. 1999), #**116**:3-hydroxy-5-cyclohexylbutyric acid (Lenz et al. 1992), #**117**:3,12-dihydroxydodecanoic acid (Lenz et al. 1992), #**118**:3,8-dihydroxy-5-cis-tetradecenoic acid (Eggink et al. 1995), #**119**:3-hydroxy-7-methoxyheptanoic acid (Kim et al. 2003), #**120**:3-hydroxy-9-methoxynon-anoic acid (Kim et al. 2003), #**121**:3-hydroxy-4-(propylthio)butyric acid (Ewering et al. 2002), #**122**:3-hydroxy-6-(propylthio)hexanoic acid (Ewering et al. 2002), #**123**:3-hydroxy-8-(propylthio)octanoic acid (Ewering et al. 2002), #**124**:3-hydroxy-4,5-epoxydecanoic acid (Eggink et al. 1995), #**125**:3-hydroxy-6,7-epoxydodecanoic acid (Lenz et al. 1992), #**126**:3-hydroxy-8,9-epoxy-5,6-*cis*-tetradecanoic acid (Lenz et al. 1992), #**127**:3-hydroxy-7-fluoroheptanoic acid (Abe et al. 1990), #**128**:3-hydroxy-9-fluorononanoic acid (Abe et al. 1990), #**129**:3-hydroxy-6-chlorohexanoic acid (Lenz et al. 1992), #**130**:3-hydroxy-8-chlorooctanoic acid (Doi and Abe 1990), #**131**:3-hydroxy-6-bromohexa-noic acid (Lenz et al. 1992), #**132**:3-hydroxy-8-bromooctanoic acid (Lenz et al. 1992), #**133**:3-hydroxy-11-bromoun-decanoic acid (Lenz et al. 1992), #**134**:3-hydroxy-2-butenoic acid (Davis 1964), #**135**:6-hydroxy-3-dodecenoic acid (Eggink et al. 1995), #**136**:3-hydroxy-2-methylbutyric acid (Satoh et al. 1992), #**137**:3-hydroxy-2-methylvaleric acid (Satoh et al. 1992), #**138**:3-hydroxy-2,6-dimethyl-5-heptenoic acid (Hazer et al. 1993)

Table 2.2 Classes of PHA synthase (adapted from Rehm 2007)

Class	Subunits	Representative species	Substrate
I	(PhaC) ~60 – 73 kDa	*Cupriavidus necator* *Sinorhizobium melioti* *Burkholderia sp.*	$3HA_{scl}$-CoA (~C3–C5) $4HA_{scl}$-CoA, $5HA_{scl}$-CoA,
II	(PhaC) ~60 – 65 kDa	*Pseudomonas aeruginosa* *P. putida*	$3HA_{mcl}$-CoA (~ $\geq$C5)
III	(PhaC)(PhaE)	*Allochromatium vinosum* *Thiocapsa pfennigii* *Synechocystis sp. PCC6803*	$3HA_{scl}$-CoA ($3HA_{mcl}$-CoA [~C6–C8], $4HA_{scl}$-CoA, $5HA_{scl}$-CoA)
IV	(PhaC)(PhaR) ~40 kDa ~22 kDa	*Bacillus megaterium* *Bacillus sp. INT005*	$3HA_{scl}$-CoA

based on their subunit composition, substrate specificity, and primary structure (Sheu and Lee 2004; Pötter and Steinbüchel 2005; Taguchi and Tsuge 2008). PHA synthases of class I and II are represented by the PHA synthase of *Cupriavidus necator* (formerly known as *Alcaligenes eutrophus* and *Ralstonia eutropha*) and *Pseudomonas aeruginosa,* respectively. The class I synthase consists of a single subunit (PhaC) with substrate preference toward scl-HA monomers. However, the PHA synthases of *Aeromonas caviae* and *Rhodospirullum rubrum* that belong to this class of enzymes do also incorporate 3HHx monomers. The class II synthases also have one subunit which actively polymerizes mcl-HA monomers (Sudesh et al. 2000; Jendrossek 2009). The class III PHA synthases is represented by *Allochromatium vinosum* and have two different subunits (PhaC/PhaE) which generally prefer to utilize scl-HA monomers (Yuan et al. 2001). PHA synthase of *Bacillus megaterium* has two subunits (PhaC/PhaR) which represent class IV of the synthase enzyme and show substrate preference similar to class III PHA synthase (McCool and Cannon 2001). The nature of the enzyme together with the kind of carbon sources fed to the microorganism and its active metabolic pathways determine the type of PHA produced (Sudesh and Doi 2005). The classes of PHA synthases are simplified in Table 2.2.

The route for P(3HB) synthesis in *C. necator* is one of the simplest and extensively studied PHA biosynthetic pathway (Fig. 2.3). Via this route, β-ketothiolase (PhaA) condenses two molecules of acetyl-CoA to form acetoacetyl-CoA. An $NADH_2$-dependent acetoacetyl-CoA reductase (PhaB) then catalyzes the conversion of acetoacetyl-CoA to (*R*)-3-hydroxybutyryl-CoA. The PhaC catalyzes the polymerization of (*R*)-3-hydroxybutyryl-CoA monomers into P(3HB) polymer (Sudesh et al. 2000; Khanna and Srivastava 2005a; Suriyamongkol et al. 2007).

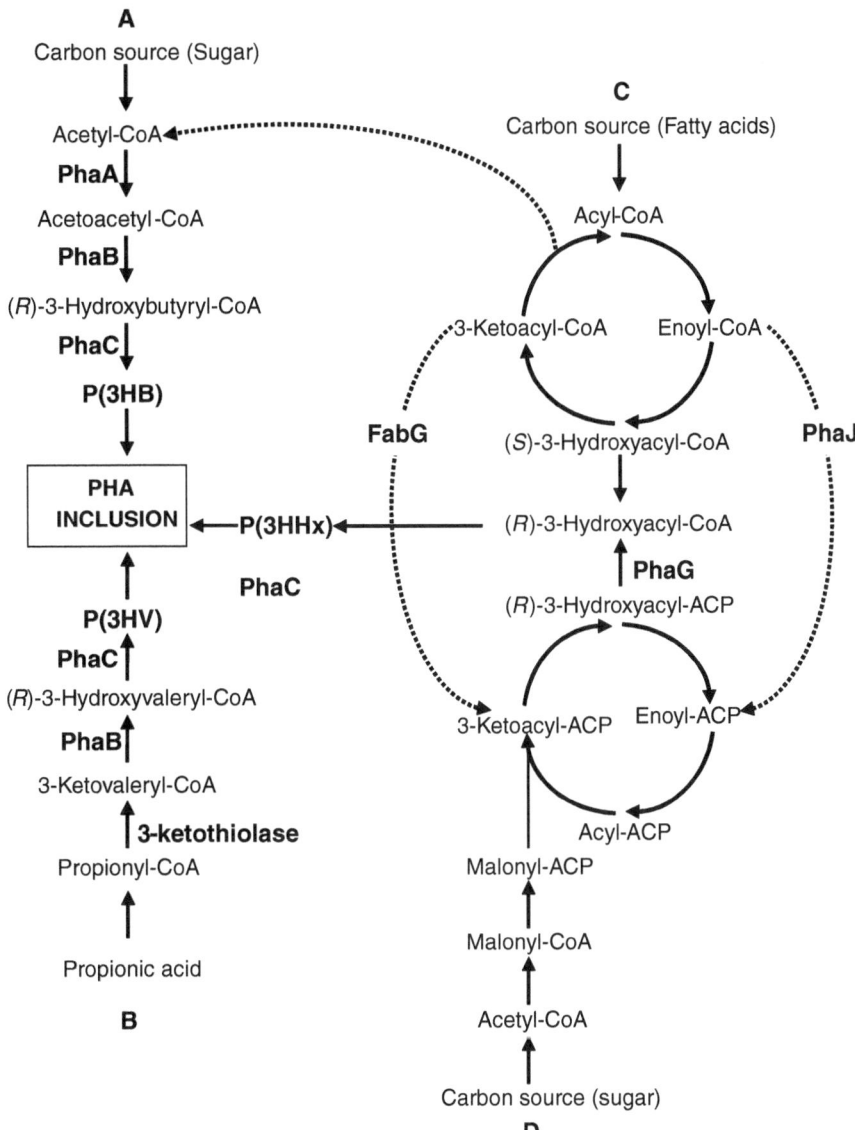

Fig. 2.3 Biosynthesis pathway of **A** P(3HB); **B** P(3HB-*co*-3HV); **C** P(3HB-*co*-3HHx) via fatty acid β–oxidation and **D** P(3HB-*co*-3HHx) via fatty acid de novo synthesis. PhaA, β-ketothiolase; PhaB, NADPH dependent acetoacetyl-CoA reductase; PhaC, PHA synthase; PhaG, 3-hydroxyl-ACP-CoA transferase; PhaJ, (*R*)-enoyl-CoA hydratase; FabG, 3-ketoacyl-CoA reductase (Sudesh et al. 2000)

According to (Anderson et al. 1990), all of the three enzymes involved in P(3HB) synthesis are to be found in the cytosol of the cell where P(3HB) accumulation occurs. Variations in the carbon sources fed to bacteria can give rise to the synthesis of PHA with different C3 to C5 monomers (Steinbüchel and

Schlegel 1991). For instance, a random copolymer composed of 3HB and 3HV [P(3HB-*co*-3HV)] can be obtained by adding propionic acid in glucose media. In this biosynthetic pathway, 3-ketothiolase mediates the condensation of propionyl-CoA with acetyl-CoA into 3-ketovaleryl-CoA which is then reduced to (*R*)-3-hydroxyvaleryl-CoA (Fig. 2.3). PhaB and PhaC involved in P(3HB) synthesis then catalyzes the polymerization of P(3HB-*co*-3HV) copolymers (Poirier 2002). The addition of aliphatic fatty acids with an odd number of carbon atoms such as valeric, heptanoic, and nonanoic acids can increase the fraction of 3HV in P(3HB-*co*-3HV) (Steinbüchel and Lütke-Eversloh 2003). *A. caviae* (Doi et al. 1995) and *A. hydrophila* (Chen et al. 2001; Han et al. 2004) are capable of naturally synthesizing P(3HB-*co*-3HHx) when fed with fatty acids of even-numbered carbons. The enoyl-CoA hydratase (PhaJ) enzyme in these strains catalyzes (*R*)-specific hydration of 2-enoyl-CoA to supply (*R*)-3-hydroxyacyl-CoAs for the polymerization of P(3HB-*co*-3HHx) synthesis via the fatty acid de novo biosynthesis or fatty acid β-oxidation pathway (Fukui and Doi 1997; Tsuge et al. 2003).

Another type of very interesting monomer constituent of PHA is 4HB. The first occurrence of 4HB in microorganisms was reported by Doi and co-workers in 1988 when *C. necator* was grown using 4-hydroxybutyric acid as the sole carbon source (Doi et al. 1988a). The same group reported that this copolymer could also be produced from different carbon substrates such as 4-chlorobutyric acid, γ-butyrolactone, and ω-alkanediols of 1,4-butanediol and 1,6-hexanediol using *C. necator* (Doi et al. 1989; Kunioka et al. 1989). Besides *C. necator*, several other wild-type microbial strains such as *Delftia acidovorans* (Saito et al. 1996), *Alcaligenes latus* (Kang et al. 1995), *Hydrogenophaga pseudoflava* (Choi et al. 1999), and *Comamonas testosteronii* (Renner et al. 1996) have also been identified to accumulate P(3HB-*co*-4HB) copolymer from the precursor substrates mentioned above. Among the precursors, 4-hydroxybutyric acid is the most suitable substrate for the synthesis of PHA with high 4HB composition.

The metabolic pathways involved in the synthesis of P(3HB-*co*-4HB) from 4-hydroxybutyric acid are shown in Fig. 2.4. Transferase or thiokinase catalyzes the conversion of 4-hydroxybutyric acid into 4HB-CoA, which is then used as the substrate by the PHA synthase in the polymerization reaction. The catabolism of 4-hydroxybutyric acid also leads to the formation of intermediates such as 3-hydroxybutyryl-CoA, resulting in the accumulation of P(3HB-*co*-4HB) copolymer. The main catabolic pathway for 4HB is probably via succinic acid semialdehyde and succinic acid pathways, which are catalyzed by 4HB dehydrogenase and succinic acid semialdehyde dehydrogenase (Valentin et al. 1995; Lutke-Eversloh and Steinbüchel 1999). All precursor substrates for the generation of 4HB monomers are first converted into 4HB-CoA, which is the immediate substrate for PHA synthase.

Another precursor substrate that is also widely used for the generation of 4HB monomers is γ-butyrolactone. The lactone is cleaved into 4-hydroxybutyric acid by the reaction of esterases or lactonases (Fig. 2.4). Subsequently, the resulting 4-hydroxybutyric acid is directly converted into 4HB-CoA either by a transferase or by a thiokinase. Again, 3HB-CoA may also be formed, thus resulting

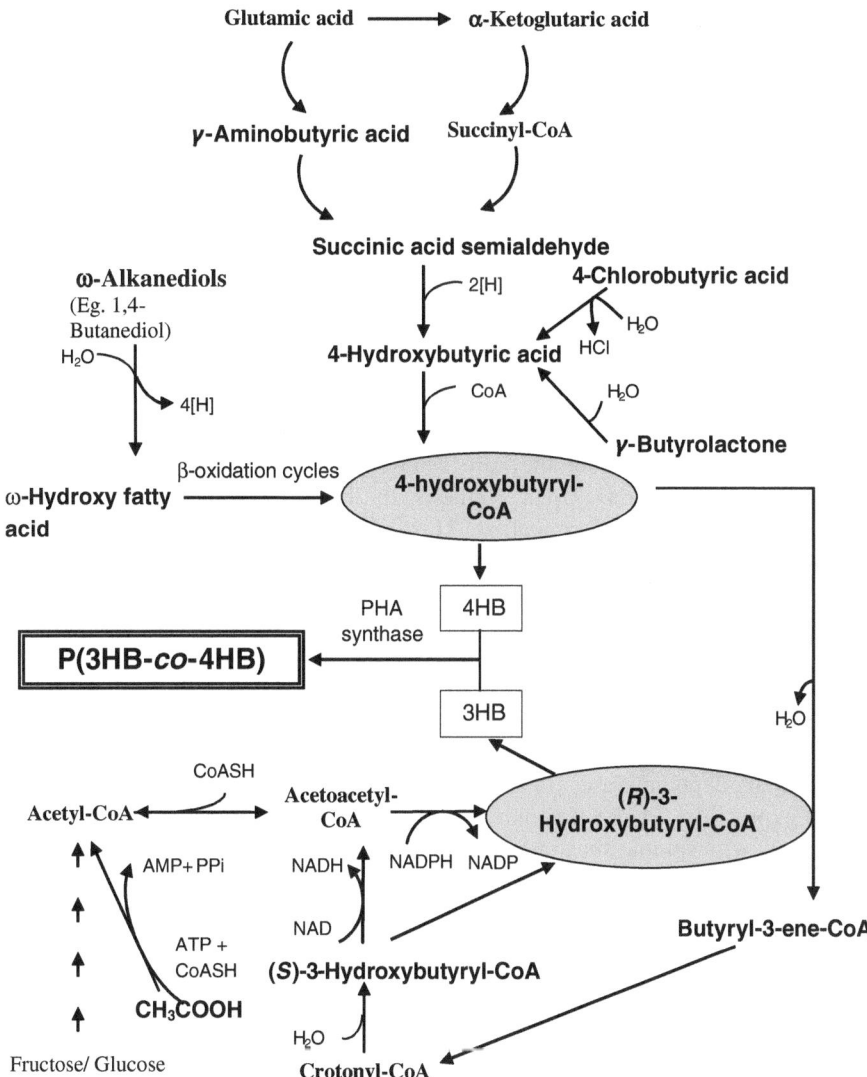

Fig. 2.4 Proposed metabolic pathways involved in P(3HB-*co*-4HB) biosynthesis from well-known 4HB precursor substrates such as 4-hydroxybutyric acid, 4-chlorobutyric acid, γ-butyrolactone, and ω-alkanediols (Doi 1990; Braunegg et al. 1998; Steinbüchel and Lütke-Eversloh 2003)

in the incorporation of 3HB monomers. When ω-alkanediols such as 1,4-butanediol, 1,6-hexanediol, 1,8-octanediol, 1,10-decanediol, and 1,12-dodecanediol are used, they are initially oxidized in two subsequent enzymatic reactions to 4-hydroxybutyric acid, which is then subsequently converted to 4HB-CoA by a transferase or a thiokinase (Fig. 2.4). In contrast to 3HB-CoA, 4HB-CoA is not a chiral intermediate. Therefore, it can be directly polymerized by the PHA synthase. Through genetic engineering, an *E. coli* transformant possessing the ability to produce P(3HB-*co*-4HB) from glucose was constructed (Valentin and Dennis 1997). This transformant expresses succinic acid semialdehyde dehydrogenase, 4-hydroxybutyric acid dehydrogenase, and 4-hydroxybutyric acid-CoA transferase gene from *Clostridium kluyveri* in addition to the PHA synthase from *C. necator*.

2.3 PHA Granule Formation

The entire process of PHA granule formation in the bacterial cell cytoplasm is an intriguing process. To begin with, the high molecular weight PHA molecules are hydrophobic but they are synthesized in an aqueous environment under ambient conditions. Upon synthesis, the bacterial cell has to ensure that the PHA molecules do not crystallize. This is because the crystalline PHA is not recognized by the intracellular PHA depolymerase. In other words, the crystalline PHA will not serve as a carbon and energy storage compound for the bacterial cell. Therefore, the synthesized PHA molecules have to be maintained in an amorphous form. The amorphous PHA granules are approximately 200–500 nm in diameter and exist as membrane enclosed inclusions (Anderson and Dawes 1990). The membrane of PHA granules is thought to comprise a phospholipid monolayer and four major granule-associated proteins consisting of PhaC, intracellular PHA depolymerases (PhaZi), phasins (PhaP), and regulator protein of the phasin expression (PhaR) (Luengo et al. 2003). The polymerization reaction of PHA starts with soluble substrate monomers and progresses to generate insoluble inclusions. In *C. necator*, PHA synthase (PhaC$_{Cn}$) becomes insoluble by granule binding because the PHA chain remains covalently linked to the enzyme during synthesis of the polymer (Gerngross et al. 1994). Phasins are the predominant proteins in the interface of PHA granules and they are considered as a class of structural proteins that play an important role in PHA biosynthesis (Pötter and Steinbüchel 2006). The existence of phasins in the phospholipid layer stabilizes the PHA granules, prevents the coalescence of the granules, and also the binding of cytosolic proteins to the hydrophobic granule surface (Steinbüchel and Valentin 1995). The level of PhaP is equivalent to the levels of P(3HB) in cell (York et al. 2001). The PhaP1$_{Cn}$ of *C. necator* is the best studied phasin (Pötter and Steinbüchel 2006). A reduction of 50 % in the total P(3HB) content was observed in a *phaP1*$_{Cn}$ gene deletion mutant of *C. necator* (York et al. 2001). Phasins influence the number and size of PHA granules (Pötter et al. 2002; Pötter and Steinbüchel 2006). Overexpression of PhaP1 was found to increase the number of granules formed (Pötter et al. 2002). However, modulation of the size and number of PHA inclusions are not only the function of PhaP, but also depends on the expression level of PhaC. The role of the phospholipid component in the PHA granule boundary is not clear.

Two models that have been proposed for the PHA granule formation in vivo are the micelle model (Gerngross et al. 1994) and budding model (Stubbe and Tian 2003). Both of these models consider the defined location of the PHA synthase. According to the micelle model (Fig. 2.5a), the PHA synthase molecules are randomly distributed in the cytoplasm and aggregate into micelle-like structures during the initial stages of 3-hydroxybutyryl-CoA polymerization. Increasing P(3HB) chain length within the micelle results in the hydrophobicity of the P(3HB)-linked synthase. Upon formation of P(3HB) chains, the polymer molecules aggregate by hydrophobic interactions and form nascent small PHA granules. The PHA granules with P(3HB)-linked synthase then disperse on the surface of emerging granules and this results in the formation of larger granules and building up of P(3HB) synthesis. The other PHA-specific surface proteins (PhaP, PhaR and PhaZi) attach to the growing surface of P(3HB) granules.

The budding model (Fig. 2.5b) assumes that the PHA synthase is associated with the cytoplasmic membrane and/or that the nascent P(3HB) chain interacts with the cytoplasmic membrane by hydrophobic interactions. This results in the formation of P(3HB) granules in between the two layers of the phospholipid bilayer. The growing P(3HB) granule is then detached from the membrane and the other PHA-specific surface proteins attach to the surface of the granules. Several studies are in agreement with this model of granule formation and reported that the emerging PHA granules tend to locate near the cell poles (Jendrossek 2005). However, contradictory results were reported by Tian et al. (2005), whereby emerging granule in wild-type *C. necator* was found to be localized close to unknown mediation elements located in the center of the cells. The granules were found to stay close to these elements until the first 24 h of accumulation. It was proposed that these mediation elements may act as nucleation sites for P(3HB) granule initiation.

2.4 Detection and Quantification of PHA

Intracellular PHA granules can be detected by a number of methods. Without any staining procedure, PHA can be observed intracellularly as light-refracting granules under phase-contrast microscope (Fig. 2.6). Rapid staining techniques using lipophilic reagents are commonly used for the detection of PHA granules, particularly during the screening process for isolation of potential PHA-producing bacteria. Lipophilic dyes such as Sudan black B (Murray et al. 1994), Nile blue A, and Nile red (Ostle and Holt 1982; Kitamura and Doi 1994; Pierce and Schroth 1994; Gorenflo et al. 1999; Spiekermann et al. 1999) are used for the staining of PHA granules in bacterial cell. The fluorescent staining of the granules using Nile Blue or Nile red is the most common detection technique in vivo (Ostle and Holt 1982). After de-staining process, cells containing lipid inclusion bodies such as PHA granules are identified by the retention of the dye. Advantage of using Nile red staining over Nile blue is that the former can be applied directly into the

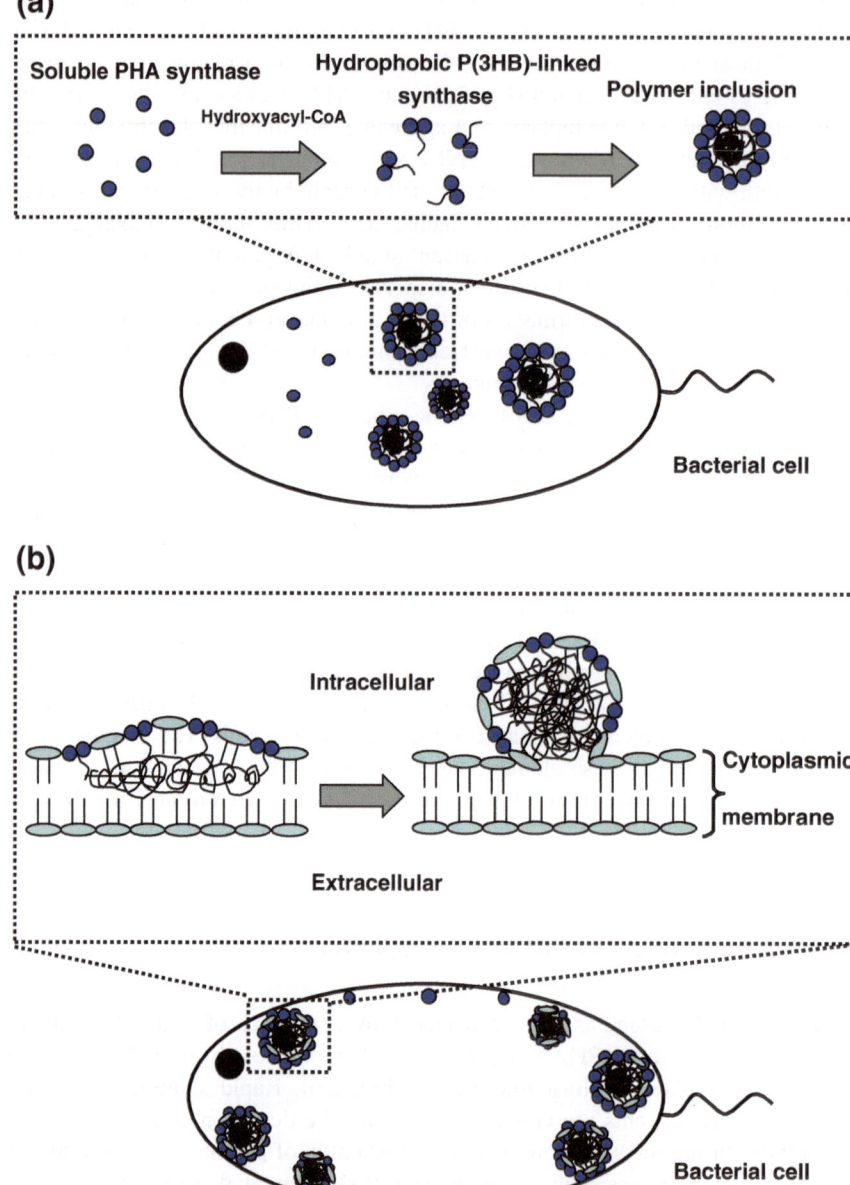

Fig. 2.5 The two models of PHA granule formation in bacteria. **a** Micelle model and **b** Budding model. The irregular lines represent the polymer chains

growing culture without causing detrimental effect on the cells. However, all these dyes are not specific toward PHA granules and can bind to any lipid compounds (Alvarez et al. 1997; Waltermann et al. 2000).

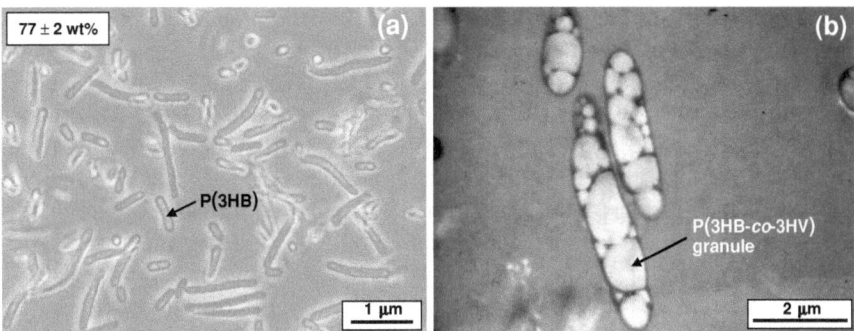

Fig. 2.6 a Under phase-contrast microscope, P(3HB) granules are observed as light-refracting granules in *C. necator* H16. P(3HB) accumulation of 77 ± 2 wt% was observed when cells were cultivated with crude palm kernel oil as the sole carbon source. **b** TEM micrographs of *C. necator* H16 cells exhibit P(3HB-*co*-3HV) inclusions as electron-transparent discrete granules. P(3HB-*co*-3HV) accumulation was observed when cells were cultivated with palm kernel oil and sodium propionate

Gas chromatography (GC) analysis involving simultaneous solvent extraction and hydrolytic esterification of PHA is generally carried out for the determination of intracellular PHA content (Braunegg et al. 1978). GC analysis with propanolysis in hydrochloric acid rather than acidic methanolysis in sulfuric acid was also reported (Riis and Mai 1988).

Quantification of microbial PHA using GC method is rapid, sensitive, reproducible, and requires only small amount of samples (5–10 mg) for the analysis. Other techniques of analysis such as IR spectrometry at 5.75 Å (Juttner et al. 1975), two-dimensional fluorescence spectroscopy, flow cytometry (Degelau et al. 1995) HPLC (Karr et al. 1983), ionic chromatography, and enzymatic determination (Hesselmann et al. 1999) were also described. For online determination of PHA content in recombinant *E. coli* system, Fourier transform mid-infrared spectrometry (FTIR) and microcalorimetric technique (Ruan et al. 2007; Jarute et al. 2004) were also reported. For precise composition determination and structural elucidation of PHA, a variety of nuclear magnetic resonance (NMR) spectroscopy techniques have also been applied and the most commonly used are proton (^{1}H) and carbon-13 (^{13}C) NMR (Doi et al. 1986; Jacob et al. 1986).

2.5 Some Commercially Attractive PHAs

2.5.1 *Poly(3-hydroxybutyrate) [P(3HB)]*

P(3HB) is the best characterized and most extensively studied of all the known PHAs. In its native granule, P(3HB) exists in amorphous (Barnard and Sanders 1989) state while extracted granules have 55–80 % crystallinity (Holmes 1988).

The weight-average molecular weight (M_w) of P(3HB) produced from wild-type bacteria is usually in the range of 1×10^4–3×10^6 g/mol. The densities of amorphous and crystalline P(3HB) are 1.18 and 1.26 g/cm^{-3}, respectively (Sudesh and Abe 2010; Doi 1990). The mechanical properties of P(3HB) in terms of Young's Modulus (3.5 GPa) and tensile strength (43 MPa) are comparable to the corresponding values of PP. However, the elongation to break of P(3HB) is only 5 % as compared to the 400 % value for PP causing the former to be more brittle and stiffer than the latter (Sudesh et al. 2000). The brittleness of the P(3HB) films is due to the formation of large crystalline domains in the form of spherulites (Sudesh and Doi 2005). Advances in genetic engineering have allowed the construction of recombinant bacteria which can produce PHA using heterologous genes. A recombinant *Escherichia coli* harboring the PHA synthase gene from *C. necator* could produce ultra-high-molecular weight (UHMW) P(3HB) with M_w values of 3×10^6–1.1×10^7 with notably improved mechanical properties (Kusaka et al. 1998). The Young's Modulus, tensile strength, and elongation to break of the UHMW homopolymer were found to be 1.1 GPa, 62 MPa, and 58 %, respectively (Kusaka et al. 1999).

P(3HB) is 100 % biodegradable and has attracted much ecological interests as it can undergo rapid degradation under environmental conditions such as aerobic, anaerobic (Nishida and Tokiwa 1993), and thermophilic conditions (Calabia and Tokiwa 2006). Therefore, it can be used to manufacture biodegradable bottles, films, adhesives, and fibers for packaging purposes. It can also be used as a raw material for enantiomerically pure chemicals and paint industry (Williams et al. 1999; Keshavarz and Roy 2010). The processing window of P(3HB) is very narrow due to its relatively close melting temperature ($T_m = 180$ °C) and thermal degradation temperature (200 °C). P(3HB) has good resistance to UV but not durable to acids and bases. It can be dissolved in chloroform or chlorinated hydrocarbons but not water; therefore, it is resistant to hydrolytic degradation (Geller 1996; Sudesh et al. 2000). Due to its biodegradable and biocompatible behavior, P(3HB)-based polymers or composites have the potential to be used in tissue engineered products such as drug release carrier, implants, or medicinal instruments (Williams and Martin 2002; Amara 2008; Heydarkhan-Hagvall et al. 2008). Low molecular weight P(3HB) has been found to be a common constituent of the animal cell membranes. The nontoxicity of P(3HB) is also further supported by the fact that a significant quantity of low M_w P(3HB) is present in the human blood and that the 3-hydroxybutyric acid is an ubiquitous metabolite in all higher living organisms (Werner and Freier 2006).

Blending of P(3HB) with other polyesters such as poly(D,L-lactide-*co*-glycolide) (PLGA) or poly(*p*-dioxanone) does not increase the flexibility of the homopolymer due to its poor miscibility (Werner and Freier 2006). In order to improve the mechanical properties of P(3HB), it is usually incorporated with other HA (3–14 carbon atom) monomers to form random copolymers in addition to increasing the molecular weight of P(3HB) (Khanna and Srivastava 2005a). However, the production cost of P(3HB) and its copolymers is high and still remains a challenge in matching that of conventional non-biodegradable plastic

(Castilho et al. 2009; Tokiwa and Ugwu 2007). Thus, various attempts are being made to utilize cheap and renewable carbon source to enable the commercialization and widespread use of these polymers.

2.5.2 Poly(3-hydroxybutyrate-co-3-hydroxyvalerate), P(3HB-co-3HV)

The commercialization of P(3HB-co-3HV) was started in 1980s under the trade name Biopol™ by Imperial Chemical Industries (ICI). Propionic acid and glucose were used to produce P(3HB-co-3HV) random copolymers containing 0–30 mol% of 3HV unit (Holmes et al. 1981). The isodimorphism of the 3HV and 3HB monomer units causes both units to co-crystallize in either of the polymer crystal lattices (Werner and Freier 2006; Cheng et al. 2008; Pan and Inoue 2009). Due to this, the degree of crystallinity of P(3HB-co-3HV) remains almost similar to P(3HB) (50–70 %) and thus, the quality of the copolymer in general is not significantly improved as compared to the homopolymer with 3HV content up to 20 mol% (Doi and Steinbüchel 2001; Khanna and Srivastava 2005a). However, the mechanical properties of P(3HB-co-3HV) strongly rely on the molar fraction of 3HV. Solution cast films of P(3HB-co-3HV) showed decrement in the value of tensile strength and Young's Modulus with an increase of 3HV fraction from 0 to 25 mol%. These results suggested that there is an increase in the flexibility of P(3HB-co-3HV) films. The toughness of the film was markedly improved as the elongation to break reached 700 % when 28 mol% of 3HV was incorporated into P(3HB). On the other hand, the lowering of T_m with increasing 3HV content (0–25 mol%) without alteration of the thermal degradation temperature provides more room for thermal processing without thermally degrading the copolymer (Sudesh and Abe 2010; Khanna and Srivastava 2005a).

2.5.3 Poly(3-hydroxybutyrate-co-3-hydroxyhexanoate), P(3HB-co-3HHx)

P(3HB-co-3HHx) possesses improved mechanical property and processability compared to P(3HB) and P(3HB-co-3HV) (Doi et al. 1995; Matsusaki et al. 2000). Co-polymerization of P(3HB) with 3HHx monomer unit which has a longer alkyl side chain avoids isodimorphism as the 3HB and 3HHx monomer units could not fit into the crystalline lattices of each other. As the 3HHx molar fraction was increased from 0 to 25 mol%, the crystallinity of P(3HB-co-3HHx) decreased from 60 to 18 %. The tensile strength of the solution-cast films of P(3HB-co-3HHx) decreased from 43 to 20 MPa while elongation to break increased from 6 to 850 % when the 3HHx content was increased from 0 to 17 mol% (Werner and Freier 2006; Cheng et al. 2008; Sudesh and Abe 2010). The nature of P(3HB-co-3HHx) copolymer becomes soft and flexible with increasing 3HHx fraction. Incorporation of small amounts of 3HHx units (5 mol%) into the 3HB sequence reduces the melting point from 179 °C to less than 155 °C (Loo et al. 2005) while further increment in 3HHx fraction up to 25 mol% results in T_m value of 52 °C. P(3HB-co-3HHx) has also been shown to be suitable candidate for blending in order to improve the ductility of stiff and brittle polyesters (Werner and Freier 2006).

2.5.4 Terpolymer

PHA that is consisted of 3 different types of monomers is classified as a terpolymer. Poly(3-hydroxybutyrate-co-3-hydroxyvalerate-co-3-hydroxyhexanoate) [P(3HB-co-3HV-co-3HHx)] produced by A. hydrophila 4AK4 transformant harboring genes phaAB is a terpolyester with higher thermal stability and elongation at "break" (this is the point where a polymer ruptures when it is stretched beyond its maximum elasticity) compared to the homopolymer P(3HB) and its copolymers P(3HB-co-3HV) or P(3HB-co-3HHx). In addition, this terpolyester had lower T_m and enthalpy of fusions (ΔH_m) than P(3HB) (Zhao and Chen 2007; Zhang et al. 2009). In another study, palm kernel oil (PKO) was used as main carbon source together with sodium propionate or sodium valerate as 3HV-precursors for the synthesis of novel compositions of P(3HB-co-3HV-co-3HHx) terpolymers which had interesting elastomeric behaviors (Bhubalan et al. 2008).

Another terpolymer poly(3-hydroxybutyrate-co-4-hydroxybutyrate-co-3-hydroxyhexanoate) [P(3HB-co-4HB-co-3HHx)] was found to have better thermal stability due to the introduction of 4HB and 3HHx monomers into P(3HB) (Xie and Chen 2008). It has lower crystallinity and better flexibility compared to P(3HB) homopolymer and copolymers. Poly(3-hydroxybutyrate-co-3-hydroxyvalerate-co-4-hydroxybutyrate) [P(3HB-co-3HV-co-4HB)] also showed superior properties over those of 3HB and its copolymer (Madden et al. 2000; Chanprateep and Kulpreecha 2006). Table 2.3 shows the comparison of different types of PHA polymer.

2.5.5 Poly(3-hydroxybutyrate-co-4-hydroxybutyrate) [P(3HB-co-4HB)]

4HB is an interesting monomer in the family of PHA. This monomer contains similar number of carbon atoms as 3HB but does not possess an alkyl side group. The linear structure of 4HB monomer suggests that 4HB could not co-crystallize into the 3HB lattice. Mitomo and co-workers had suggested that 4HB lattice had to undergo deformation process in order to accommodate into the 3HB monomer lattice (Mitomo et al. 2001). The crystallinity of P(3HB-co-4HB) copolymer decreased from 60 to 15 % when the 4HB monomer composition was increased

Table 2.3 Properties of some P(3HB) homopolymer, copolymers and terpolymers

Polymer	T_m^a (°C)	T_g^b (°C)	Tensile strength (MPa)	Elongation to break (%)	Reference
P(3HB)	177	4	43	5	(Tsuge 2002)
P(3HB-co-20 mol% 3HV)	145	−1	20	50	(Tsuge 2002)
P(3HB-co-16 mol% 4HB)	150	−7	26	444	(Tsuge 2002)
(3HB-co-10 mol% 3HHx)	127	−1	21	400	(Tsuge 2002)
P(3HB-co-22 mol% 3H4MV)	126	−2	11	380	(Tanadchangsaeng et al. 2009)
P(3HB-co-3 mol% 3HV-co-93 mol% 4HB)	55	−52	14	430	(Chanprateep and Kulpreecha 2006)
P(3HB-co-7 mol% 4HB-co-20 mol% 3HHx)	–	−8	1	364	(Xie and Chen 2008)
P(3HB-co-2 mol% 3HV-co-7 mol% 3HHx)	144	−3	22	312	(Bhubalan et al. 2008)

[a]Melting temperature
[b]Glass-transition temperature

Table 2.4 Physical and mechanical properties of P(3HB-co-4HB) copolymers (Saito et al. 1996)

Properties	4HB composition (mol%)						
	0	16	64	78	82	90	100
T_m^a	178	130	50	49	52	50	53
T_g^b	4	−7	−35	−37	−39	−42	−48
Crystallinity (%)	60	45	15	17	18	28	34
Tensile strength (MPa)	43	26	17	42	58	65	104
Elongation at break (%)	5	444	591	1120	1320	1080	1000

[a]Melting temperature
[b]Glass transition temperature

from 0 to 64 mol%. Nevertheless, an increase in polymer crystallinity from 15 to 34 % was detected with copolymers containing higher 4HB monomer composition of 78–100 mol%.

P(3HB-co-4HB) copolymers possess lower T_m compared to P(3HB) homopolymer. The T_m ranges from 178 to 50 °C with increasing 4HB monomer composition of 0–100 mol% (Table 2.4). The T_g also showed a similar declining trend from 4 to −48 °C when the 4HB content increases. P(3HB-co-4HB) copolymers containing high 4HB monomer composition exhibits elastomeric property. The elongation to break values of P(3HB-co-4HB) copolymers were in the range of 5–1320 % with increasing 4HB content (Saito et al. 1996). P(3HB-co-82 mol% 4HB) was found to be a very flexible polymer with an elongation to break of up to 1320 %. The tensile strength of these copolymers showed a declining trend from 43 to 17 MPa as the 4HB monomer composition increased from 0 to 64 mol%. However, as the 4HB monomer composition was further increased from 64 to 100 mol%, an increase in tensile strength from 17 to 104 MPa was recorded.

P(3HB-*co*-4HB) copolymers are known to be biocompatible material. The bio-compatible nature of these copolymers allows it to be utilized for various medical applications. These PHA polymers have been tested in tissue engineering applications as surgical sutures, bone plates, implants, gauzes, osteosynthetic materials, and also as matrix material assisting slow release of drugs and hormones (Zinn et al. 2001; Williams and Martin 2002; Sudesh 2004; Chen and Wu 2005; Freier 2006). Recently, electrospun nanofibers of P(3HB-*co*-4HB) have been evaluated as scaffolds in vivo and in vitro (Ying et al. 2008).

2.5.6 PHA with Unusual HA Monomers

An increasing number of new monomers are found to be accepted as substrates for the PHA synthase (Fig. 2.2). PHA with different functional groups in the side chain such as halogens, carboxyl, hydroxyl, epoxy, phenoxy, cyanophenoxy, nitro-phenoxy, thiophenoxy, and methylester groups were synthesized using bacteria possessing different metabolic pathways (Kessler et al. 2001). The PHA polymer containing 3-hydroxy-5-phenylvalerate had been produced by *P. oleovorans* from mixtures of 5-phenylvaleric acid with either *n*-nonanoic acid or *n*-octanoic acid. Up to 40 mol% of this phenyl containing monomer coupled with 32 wt% PHA content were obtained (Kim et al. 1991). When 11-cyanoundecanoic acid and *n*-nonanoic acid were fed as the carbon feedstock to *P. oleovorans*, a polymer consisting of 32 mol% of cyano-containing monomers (mainly 9-cyano-3-hydroxyno-nanoate and 7-cyano-3-hydroxyheptanoate) was synthesized (Lenz et al. 1992).

The biosynthesis of PHA containing halogenated functional group by *P. oleovorans* was first reported in 1990 by Doi and coworkers. A random PHA copolymer containing about 38 mol% of this functional group was obtained when equimolar of nonanoic acid and 11-bromoundecanoic acid was supplemented to the culture medium. In addition, mixtures of nonanoic acid or octanoic acid with 6-bro-moundecanoic acid and 8-bromo-octanoic acid have also been identified as potential precursors for the production of PHA containing monomer with halogenated functional group by *P. oleovorans* (Abe et al. 1990; Doi and Abe 1990; Kim et al. 1992; Kim et al. 1996).

PHA containing sulfur atom could also be synthesized by feeding the polythioesters-accumulating bacteria such as *C. necator* with special mercap-toalkanoic acid such as 3-mercaptopropionic acid, 3,3'-dithiodipropionic acid, and 3-mercaptobutyric acid (Lutke-Eversloh et al. 2001; Lutke-Eversloh et al. 2002). The PHA produced from these mercaptoalkanoic acids is usually heter-opolymers which contain oxoester and thioester. A genetically engineered *E. coli* transformant harboring the butyrate kinase and phosphotransbutyrylase genes of *Clostridium acetobutylicum* as well as the *phaC* of *Thiococcus pfennigii* had shown to have the ability to produce homopolymer of poly(3-mercaptopropionate), poly(3-mercaptobutyrate), and poly(3-mercaptovalerate) (Kawada et al. 2003; Kim et al. 2005; Lutke-Eversloh et al. 2002). The studies above proved that PhaC is also capable of incorporating non-natural substrates such as mercaptoalkanoic acid for the production of novel polymers. Nevertheless, these polythioesters could not be degraded like normal PHA due to the presence of sulfur group (oxoester bonds) in the structure which is not recognized by the PHA depolymerase (Yu et al. 2007).

2.5.7   Recent Discovery of LA-Based Monomers

Recently, it has been demonstrated that the PHA synthase has the ability to synthesize PLA (Taguchi et al. 2008; Park et al. 2008a, b, c). In an interesting study done by Taguchi and coworkers, the evolved PHA synthase of *Pseudomonas* sp. 61-3 was found to co-polymerize 3HB-CoA and lactate-CoA (LA-CoA) into P(3HB-*co*-LA) via microbial process (Taguchi et al. 2008). In current industrial process, PLA is polymerized via chemo-processes (metal-catalysts) from lactic acid (2-hydroxypropionic acid) (Enomoto et al. 1994). Lactic acid could be derived from anaerobic fermentation of sugars and starch but could not be directly polymerized into PLA. Since the PHA monomer shares similar hydroxyl acid group with lactic acid, Taguchi and coworkers genetically modified and constructed an *E. coli* transformant capable of converting lactic acid into LA-CoA which is then polymerized by a mutant synthase of *Pseudomonas* sp. 61-3 into lactate polymer in vivo. This breakthrough has further confirmed that the PHA synthase is an interesting enzyme which possesses a wide range of substrate specificity. In addition, this study also described a one-step process for producing LA-based polymers using microbial system.

2.6   Production of PHA in Microorganisms and Plants

Unlike the chemically synthesized plastics, synthesis of PHA involves a complex biological system in which several factors inherent to the microorganisms must be seriously weighed to optimize the PHA accumulation. These determining factors include the ratio of carbon source to nitrogen source, the presence of macro elements (magnesium, potassium, oxygen, phosphate, and iron), and suitable culture conditions (time, temperature and pH) (Lee et al. 2004; Kadouri et al. 2005). In most cases, the presence of excess carbon source is a prerequisite to initiate the accumulation of PHA, which is usually initiated by the limitation of certain nutrients such as nitrogen or phosphorous. Generally, higher concentrations of carbon sources result in higher pool of HA monomers, hence leading to more PHA accumulation.

Selection of suitable carbon substrates is also another critical factor that determines the overall performance of the fermentation process as well as significantly influencing the cost of the final product. Therefore, the simplest approach

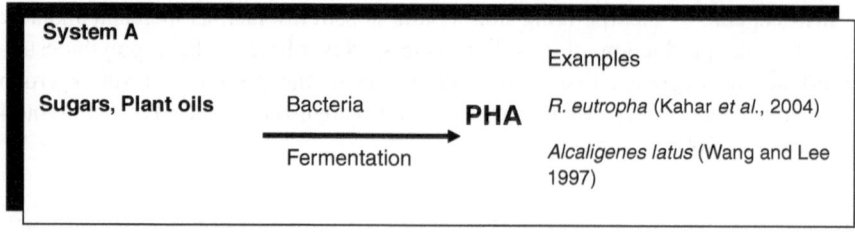

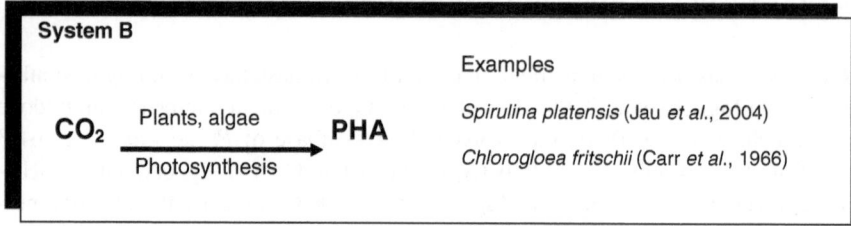

Fig. 2.7 Strategies for the production of PHAs using different renewable resources as the starting material

is to choose the cheapest and most readily available carbon substrates (preferably bio-based) that could support the microbial growth efficiently. Sugars (Borah et al. 2002), plant oils (Loo et al. 2005), and several agricultural by-products such as beet molasses (Omar et al. 2001), alphechin wastes (Pozo et al. 2002), and starch are among the various relatively cheap renewable resources that are being studied in detail at this moment. Figure 2.7 shows the two general strategies for the synthesis of PHA based on different processes (bacterial fermentation or photosynthesis) of utilizing the carbon sources. At present, studies on the biosynthesis of PHA mostly employ the first strategy, i.e., the microbial conversion of reduced carbon sources into PHA via fermentation processes.

PHA is produced by different bacterial strains. One of the most studied strain is *C. necator* (formerly known as *Wautersia eutropha, Ralstonia eutropha* or *Alcaligene eutrophus*). It was used in industrial production by Imperial Chemical Industries (ICI PLC) to produce P(3HB-*co*-3HV) under the trade name of Biopol™. The Biopol™ patents have now been acquired by Metabolix Inc. (USA) (Verlinden et al. 2007). Until now, *C. necator* is still being used widely for bacterial fermentation as it is an efficient strain. Other important strains that have been studied for PHA production are *Bacillus* spp., *Alcaligenes* spp., *Pseudomonas* spp., *Aeromonas hydrophila, Rhodopseudomonas palustris,* recombinant *Escherichia coli, Burkholderia sacchari* , and *Halomonas boliviensis* (Verlinden et al. 2007).

C. necator is the model bacterium for the biosynthesis of PHA. This strain generally initiates the synthesis of PHA when either nitrogen or phosphorous is limited during growth (Kahar et al. 2004). A similar phenomenon occurs in several other PHA producers including *Burkholderia cepacia* (Zazali and Tan 2005; Mitomo et al. 1999), *Pseudomonas* sp. (Choi et al. 2003), and *A. hydrophila*

(Qiu 2006). *Alcaligenes latus* is an excellent P(3HB) producer because it grows rapidly on sucrose (Tamer et al. 1998) and beet molasses (Hangii 1990) and accumulates P(3HB) almost immediately during active growth without nutrient limitation. *A. latus* DSM 1123 takes approximately 5 h to achieve the accumulation of P(3HB) up to 80 wt% of the dry cell weight (Wang and Lee 1997). Spore-forming *Bacillus* strains are able to produce novel terpolymer (Łabu·zek and Radecka 2001). However, the environmental conditions favorable for PHA productions also encourage spore production. Hence, these two metabolic processes may interfere with one another resulting in the reduction of PHA production. Therefore, nonspore-forming mutant strains of *Bacillus* are being evaluated for their PHA production ability.

Genetic engineering is a useful tool to design strains that have better ability to produce PHA with tailored properties (Verlinden et al. 2007). Most wild-type PHA-producing bacteria have a relatively slow growth rate during fermentation. In addition, the cells are also usually hard to lyse, making PHA extraction and purification difficult. The wild-type PHA producers also contain metabolic pathways for PHA degradation and utilization. In short, some of these wild-type bacteria are unsuitable for PHA production at industrial scale.

The commonly used strains for genetic recombination approach are *E. coli*, *C. necator* PHB⁻4, and *Pseudomonas*. Bacteria such as *E. coli* are incapable of producing or degrading PHA. However, it can achieve high cell density within a short period of time even at 37 °C. Fast growth will allow it to maximize the amount and extent of its polymer accumulation. *E. coli* cells are also easy to lyse (Steinbüchel and Schlegel 1991; Madison and Huisman 1999). Furthermore, *E. coli* can utilize various carbon sources, e.g., glucose and lactose, as well as inexpensive carbon sources such as whey, molasses, and hemicellulose hydrolysate (Lee et al. 1995). Hence, *E. coli* is a viable alternative host to generate higher PHA yield.

Two main challenges in using recombinant *E. coli* for PHA production are the instability of the introduced PHA biosynthetic genes and loss of plasmid due to metabolic loads (Lee et al. 1994; Madison and Huisman 1999). In order to maximize PHA production, recombinant *E. coli* strains harboring the *C. necator* PHA synthesis genes in a stable high copy number plasmid have been developed and used (Zhang et al. 1994; Lee et al. 1994). Recombinant *E. coli* cultured under optimal conditions has been shown to accumulate P(3HB) up to 85 % of the dry cell weight (Zhang et al. 1994). There are two strategies in developing a recombinant bacterium that is capable of producing PHAs from cheap carbon sources. The first approach is the introduction of substrate utilization genes into a natural PHA producer. The second approach is to introduce PHA biosynthesis genes into a non-PHA-producing strain, which is able to use renewable carbon sources efficiently (Lee 1996b).

The approach of producing P(3HB) from CO_2 is probably the ideal way of maintaining the carbon balance in the ecosystem (System B in Fig. 2.7). The idea is commercially attractive because it is based on the concept of sustainability and the utilization of the cheapest renewable source as the starting material. Higher plants naturally could not accumulate P(3HB) due to the absence of key enzymes involved in the biosynthesis of PHA. However, it is now technologically possible to produce P(3HB) in transgenic plants such as *Arabidopsis thaliana, Gossypium hirsutum* (cotton), and *Zea mays* (corn) (Poirier et al. 1992a, b; Hahn 1995) whereby the polymers are accumulated in the plant cell cytoplasm (Bohmert et al. 2002). It is well established that acetyl-CoA is the intermediate substrate for the synthesis of 3HB monomers. Since acetyl-CoA is present in cytosol, plastid, mitochondrion, and peroxisome of plant cells, it was perceived that synthesis of P(3HB) is possible in these compartments via heterologous expression. The production of P(3HB) in *A. thaliana* was initiated via transformation of *A. thaliana* with the PHA synthase gene of *C. necator* to induce polymerization of the intermediates derived from the plant's endogenous 3-ketothiolase and acetoacetyl-CoA reductase. Several other attempts to produce higher amounts of PHA in a cost-effective manner via transgenic plants are currently in progress (Matsumoto et al. 2005; Matsumoto et al. 2006; Nielsen 2007). It is believed that transgenic plants could be developed into commercially viable PHA production systems in the future.

The approach of using algae and cyanobacteria are advantageous in many ways. Algae and cyanobacteria are ecologically important to maintain the balance of atmospheric level of carbon dioxide. Cyanobacteria such as *Spirulina platensis* is biotechnologically important due to its high nutritional value (Ciferri and Tiboni 1985), which suggests that the residual cell biomass following purification of PHA is feasible as animal feed (Belay et al. 1996). In spite of highly optimized PHA accumulation from bacterial fermentation (80 wt% PHA of cell dry weight), this process requires expensive continuous rich oxygen supply, which eventually leads to higher feeding and operation cost. On the other hand, producing PHA directly from plants would require the least amount of energy, making it the most efficient process.

Like other higher plants, cyanobacteria are photoautotrophs. Some cyanobacteria possess enzymes involved in the biosynthesis of PHA. Strains such as *Oscillatoria limosa, S. platensis,* and *Synechocystis* sp. are considered to be of interest due to their natural ability to accumulate PHA as a storage product of CO_2 fixation. Generally, the PHA content found in cyanobacteria is quite low (6–14 wt% of their dry weight). Asada et al. (1999) have reported that up to 27 wt% P(3HB) of dry cell weight was accumulated when *Synechococcus* MA19 was cultivated under photoautotrophic and nitrogen starved conditions, which is the highest among oxygenic photosynthetic organisms. The apparent low yield of P(3HB) compared to bacteria is probably due to several reasons; (i) the larger size of cyanobacterial cells compared to bacterial cells, (ii) thicker cell wall that could prevent efficient water removal during freeze-drying. These possibilities might contribute to the higher dry mass than the actual mass, thus contributing to the lower P(3HB) content as determined by GC (Sudesh et al. 2001).

Electron microscopy has been employed to identify the locations, sizes, and number of granules in cyanobacteria. This technique revealed interesting observation whereby the number and sizes of PHA granules were comparable to that of some PHA-producing bacteria (Sudesh et al. 2001). In addition, freeze-fracture electron microscopy observation on *Synechocystis* sp PCC 6803 showed that the internal microstructure of P(3HB) was essentially similar to the P(3HB) granules found in typical bacteria (Sudesh et al. 2000). During active growth of *S. platensis* UMACC 161, no apparent P(3HB) granules were detected (Jau et al. 2005). When the cells were transferred into nitrogen-free conditions, many PHA granules could be observed as electron-transparent inclusions in the cytoplasm. These granules were surrounded by the thylakoid membrane and could be easily distinguished from other inclusions in the cytoplasm.

Bacterial synthesis of PHA is normally carried out in batch or fed-batch fermentations. To initiate PHA accumulation, the cultivation medium is prepared in such a way that one nutrient (generally nitrogen or phosphate) limits the growth of bacterial cells. However, the carbon source is supplied in excess. The depletion of the selected nutrient acts as a trigger for the metabolic shift to PHA biosynthesis. For example, under balanced growth in *C. necator*, carbohydrate is catabolized via Entner-Doudoroff pathway to pyruvate which is then converted through dehydrogenation to acetyl-CoA. Subsequently, the acetyl-CoA enters the TCA cycle and is utilized for reproductive growth. However, in the occurrence of certain nutrient limitation, cessation of protein synthesis results in high concentrations of NADH and NADPH which inhibit citrate synthase and isocitrate dehydrogenase (Braunegg et al. 1998). Citrate synthase regulates the P(3HB) production by controlling carbon flux into TCA cycle. Inhibition of citrate synthase slows down the TCA cycle and channels acetyl-CoA toward P(3HB) biosynthesis (Dawes and Senior 1973).

Two different approaches have been developed in batch cultivation, which are: one-stage cultivation and two-stage cultivation. In one-stage cultivation, growth of cells and PHA accumulation occurs simultaneously. While two-stage cultivation consists of a cell growth phase which is carried out in a separate nutrient enriched medium. The cells are then transferred into a nutrient limiting medium for PHA accumulation phase. Usually, the cultivation period ranges from 24 to 96 h. During the period of cultivation, cells undergo a sequence of growth phases, such as lag phase, exponential phase, and PHA production phase. In a fed-batch culture, the cells are continuously fed with selected carbon source after they have entered the late exponential phase. Normally, large-scale or industrial-scale production systems use fed-batch cultivation mode (Kahar et al. 2004; Chen et al. 2001). The fed-batch method generally yields high cell densities which consequently reduces the overall production cost (Lee and Choi 1998, Kellerhals et al. 1999a, b). PHA concentrations of greater than 80 g/L with productivity greater than 2 g PHA/L/h can be routinely produced by fed-batch cultivation (Braunegg et al. 1998). Other modes of cultivation have also been evaluated, such as: pH–stat based cultivation whereby carbon source is fed based on the fluctuation of pH (Choi and Lee 1999) and chemo-stat method whereby culture medium is continuously exchanged with sterile growth medium (Ren et al. 2007). For a successful PHA production system at commercial scale, it is essential to investigate many physical and biological parameters (e.g., optimal temperature, pH, substrate and growth medium) in order to establish optimal fermentation conditions.

PHA production by microbial fermentation is perfectly integrated into nature's closed cycle of carbon attributable to its renewable nature (Braunegg et al. 2004; Sudesh and Iwata 2008). PHA synthesis is based on renewable resources;

agricultural products such as sugars and fatty acids are commonly used as the carbon and energy sources (Solaiman et al. 2006; Kadouri et al. 2005). These agricultural feedstocks are derived from carbon dioxide and water. After their bioconversion to biodegradable PHA and the usage periods, their final oxidative breakdown merely yields CO_2 and water.

2.7 Carbon Sources for PHA Biosynthesis

When there is a lack of growth nutrients except carbon source in bacterial cells, the surplus carbon is metabolized via a few metabolic pathways to produce PHA. The type of the carbon sources supplied greatly affects the biosynthesis of PHA. The carbon sources can be categorized into two groups, namely, structurally related carbon source and structurally unrelated carbon source (Taguchi et al. 2004). The former group refers to the carbon sources that result in HA monomers that have similar chemical structures to them. The latter group consists of carbon sources that generate HA monomers which have completely unrelated chemical structure to them. According to previous studies, the cost of carbon substrate is estimated to contribute approximately 28–50 % of the total production cost (Lee and Choi 1998; Braunegg et al. 2004). Regardless of the type of carbon source used and monomers produced, the concern on the production cost is always affecting the wide use of PHA. Therefore, much efforts is ongoing to reduce the production cost in order to make PHA economically feasible.

2.7.1 Sugars

In most PHA production studies, sugars such as glucose (Łabużek and Radecka 2001; Valappil et al. 2007), gluconate (Hoffmann and Rehm 2004; Valappil et al. 2007), sucrose (Zhang et al. 1994; Valappil et al. 2007), and fructose (Tsuge et al. 2005; Valappil et al. 2007) are usually used. Initial large-scale production of PHA was also carried out using sugars as the carbon source. The industrial-scale production of PHA using 200,000-L stirred fermentation vessels began in the 1970s by ICI, in Great Britain. *C. necator* was chosen for the production of P(3HB) by using fructose as the sole carbon source (Byrom 1987). With co-feeding of propionate, a glucose-utilizing mutant strain of *C. necator* (ATCC 11599) was employed by Monsanto to synthesize P(3HB-*co*-3HV), a copolymer with better flexibility and impact resistance than P(3HB). The first large-scale production of P(3HB-*co*-3HHx) copolymer from glucose was performed using *Aeromonas hydrophila* 4AK4 in 20,000-L fermentor (Chen et al. 2001). Sucrose was also used as the sole carbon source by Chemie Linz GmbH (Linz Austria) for the production of P(3HB) at up to 1,000 kg per week by employing *Alcaligenes latus* (Hrabak 1992). However, the high cost of these sugars makes PHA production cost incomparable to that of the conventional plastics. In order to reduce the production cost of PHA, the use of reasonably priced carbon sources such as plant oils or fatty acids, wastes from agricultural or food industries and even carbon dioxide have been studied (Tsuge 2002).

2.7.2   Triglycerides

For an economically feasible production of PHAs, there have been a considerable amount of interests in using inexpensive carbon substrates as an alternative to sugars (Lee and Choi 1998; Lee et al. 1999; Tsuge 2002; Solaiman et al. 2006). Various plant oils and their derived fatty acids have emerged as superior candidates for PHA production by various PHA producers. PHA accumulation of more than 80 wt% of DCW was reported when PKO, corn oil, coconut oil, and olive oil was supplemented as the sole carbon source. Besides, a range of commercially important vegetable oils were also found to be feasible for the bioconversion to various types of PHA (Fukui and Doi 1998). Vegetable oils, which are composed of triglycerides with long-chain fatty acids, have higher carbon content per unit weight as compared to sugars, which gives a higher theoretical yield coefficient; over 1 g PHA per 1 g of vegetable oil used (Akiyama et al. 2003). According to Akiyama et al. (2003), plant oils yield approximately twofold higher amounts of P(3HB) (0.6–0.8 g/g) as compared to that with glucose (0.3–0.4 g/g) which could reduce the overall PHA production cost and energy consumption. This finding gave rise to the exploitation of plant oils such as soybean oil, corn oil, and palm oil as carbon substrates for PHA production. Soybean oil (Tsuge et al. 2004), palm oil products namely crude palm kernel oil (CPKO) and PKO (Loo et al. 2005; Bhubalan et al. 2008; Kek et al. 2008), spent palm oil (Rao et al. 2010), and hydrolyzed corn oil (Shang et al. 2008) have been studied and found to be excellent carbon sources for higher cell biomass and increased PHA yield. Volatile fatty acids (VFAs) (Kourmentza et al. 2009) have also been used in mixed culture. It led to better final yields of PHA/VFA and more rapid PHA accumulation rates.

2.7.3   Industrial Waste Stream and By-products

The idea of using various agricultural and industrial waste streams or by-products as the fermentative substrate for PHA synthesis has gained interest among researchers due to their availability and renewable merits. There have been several reports on the production of P(3HB) from waste or residual materials of agro-industrial sector by wild-type P(3HB) producers. Cheese whey, xylose, molasses, and bagasse/starch hydrolysate are among the inexpensive carbon substrates studied for biosynthesis of P(3HB) (Pijuan et al. 2009; Rodgers and Wu 2010; Chen et al. 2006; Gouda et al. 2001; Lee and Choi 1998). Other oily substrates such as tallow, waste frying oil, and its derived fatty acids were also found to be feasible for the bioconversion into PHA (Fernandez et al. 2005; Taniguchi et al. 2003). Bioconversion of renewable and cheap waste/by-products can offer multiple benefits to the environment by diverting high pollution potential products to the synthesis of value-added green materials. In other words, the production costs of PHA could be reduced and sludge generation could be minimized at the same time. Xu et al. (2010) studied the feasibility of using wheat-based biorefinery waste for the production of P(3HB). In another example, the use of combinations of dairy waste, rice bran, and seawater for the production of P(3HB) was carried out (Pandian et al. 2010). In addition to the aforementioned examples, milk and ice-cream processing wastewater (Chakravarty et al. 2010), fermented sugar cane molasses (Bengtsson et al. 2010), waste glycerol (Cavalheiro et al. 2009), starch (Yu 2001), triacylglycerols (Solaiman et al. 2002), and methanol (Mokhtari-Hosseini et al. 2009) were also found to be economically competitive carbon sources for the production of PHA. Besides these common industrial wastes, organic acids such as lactic acid, acetic acid, and propionic acid, which are by-products of the anaerobic fermentation processes had also been evaluated for P(3HB) and P(3HB-co-3HV) synthesis (Tsuge et al. 1999; Kobayashi et al. 2000; Tsuge et al. 2001). Hassan and co-workers used mixtures of organic acids derived from anaerobically treated palm oil mill for PHA production (Hassan et al. 1997b). Utilization of these carbon sources for large-scale PHA production could potentially reduce the production cost and minimize waste.

2.8 Extracellular PHA Degradation

Extracellular PHA degradation is the degradation of PHA that is found outside of the cell. When cells lyse, the intracellular PHA granules may be released from the cell cytoplasm. Native intracellular PHA granules are in the amorphous state, whereas the extracellular PHA is partially crystalline. These extracellular PHA granules become available as a carbon source to other bacterial cells. In order to use the PHA as a carbon source, enzymes that are able to hydrolyze PHA into its monomeric units are needed. Many types of bacteria, including some that do not have the ability to synthesize PHA, are known to secrete extracellular PHA depolymerases, which are carboxyesterases (Jendrossek and Handrick 2002). Microorganisms need to excrete extracellular PHA depolymerase because solid PHA polymer of high molecular weight is unable to diffuse through the cell wall of bacteria (Abe et al. 2001). PHA depolymerases hydrolyze water-insoluble PHA into water-soluble monomers and/or oligomers which are then assimilated and metabolized within cells into CO_2 and H_2O (Sridewi et al. 2006). The rate of biodegradation was found to be influenced by several factors in a given environment such as microbial population, temperature, moisture level, pH, and nutrient supply besides the composition, crystallinity, additives, and surface area of PHA itself (Abou-Zeid et al. 2001).

Extracellular PHA depolymerases are ubiquitous in the environment (Tokiwa and Calabia 2004). The earliest discovery of PHA-degrading bacteria belongs to

Bacillus, *Pseudomonas*, and *Streptomyces* (Chowdhury 1963). From then onwards, a vast number of aerobic and anaerobic PHA-degrading bacteria have been isolated from various environments such as soil, sludge, and seawater. Mergaert et al. (1993) reported the isolation and identification of 295 strains capable of degrading PHB and P(3HB-*co*-3HV) copolymer from soils. Recently, a thermoalkanophilic P(3HB-*co*-3HV) esterase was reported to be produced by the soil isolate *Streptomyces* sp. IN1 (Allen et al. 2011). An extracellular mcl-PHA depolymerase which was purified from *Thermus thermophilus* HB8 has the ability to hydrolyze mcl-PHA and *p*-nitrophenyl (*p*NP) esters but not scl PHA (Papaneophytou et al. 2011). Volova and coworkers (2010) isolated several PHA-degrading strains identified as *Enterobacter* sp., *Bacillus* sp., and *Gracilibacillus* while studying the biodegradability of P(3HB) and P(3HB-*co*-11 mol% 3HV) in a tropical marine environment of South China Sea (Volova et al. 2010). Many P(3HB)-degrading fungi have also been identified thus far (Matavulj and Molitoris 1992; Sang et al. 2001; Shah et al. 2010). Various microorganisms were found to colonize the surface of P(3HB), P(3HB-*co*-5 mol% 3HV) and P(3HB-*co*-5 mol% 3HHx) cast films after four weeks of incubation in a tropical mangrove sediment (Fig. 2.8).

Degradation of PHA films proceeds via surface erosion mechanisms whereby bacteria attach on the porous area on the film surface and secrete depolymerase enzymes to perform catalytic action on the polymer (Wang et al. 2004). These enzymes go through a two-step heterogeneous enzymatic hydrolysis to disintegrate PHA films which includes the adsorption of enzyme on the surface of PHA films by the substrate-binding domain of the depolymerase followed by the hydrolysis of PHA molecule by the catalytic domain (Feng et al. 2004). The catalytic domain contains the catalytic machinery composed of a catalytic triad (Ser-His-Asp). The serine is part of a lipase box pentapeptide Gly-X-Ser-X-Gly, which has been found in all known hydrolases such as lipases, esterases, and serine proteases. A linker region connects the aforementioned two domains (Jaeger et al. 1994). The degradation of different PHA films depends greatly on the specificity of the active site in the catalytic amino acid domain structure of a depolymerase enzyme (Kasuya et al. 1999; Shinomiya et al. 1998). It is also well documented that although this enzyme is able to adhere on various substrate surfaces, but the active site in its catalytic domain is specific for the hydrolysis of PHA molecules (Kasuya et al. 1996).

2.8.1 Lipolytic Enzymes

In order to carry out PHA production from triglycerides, the cells need to hydrolyze the triglycerides into free fatty acids that can be transported into the cells. For this, the cells need to secrete an enzyme called lipase. A wide variety of organisms is known to secrete lipolytic enzymes, mainly for lipid metabolism and signal transduction. Lipolytic enzymes can be classified into different classes, including lipases, esterases, and phospholipases (Arpigny and Jaeger 1999). Esterases are enzymes that hydrolyze ester bonds of soluble or partially soluble molecules,

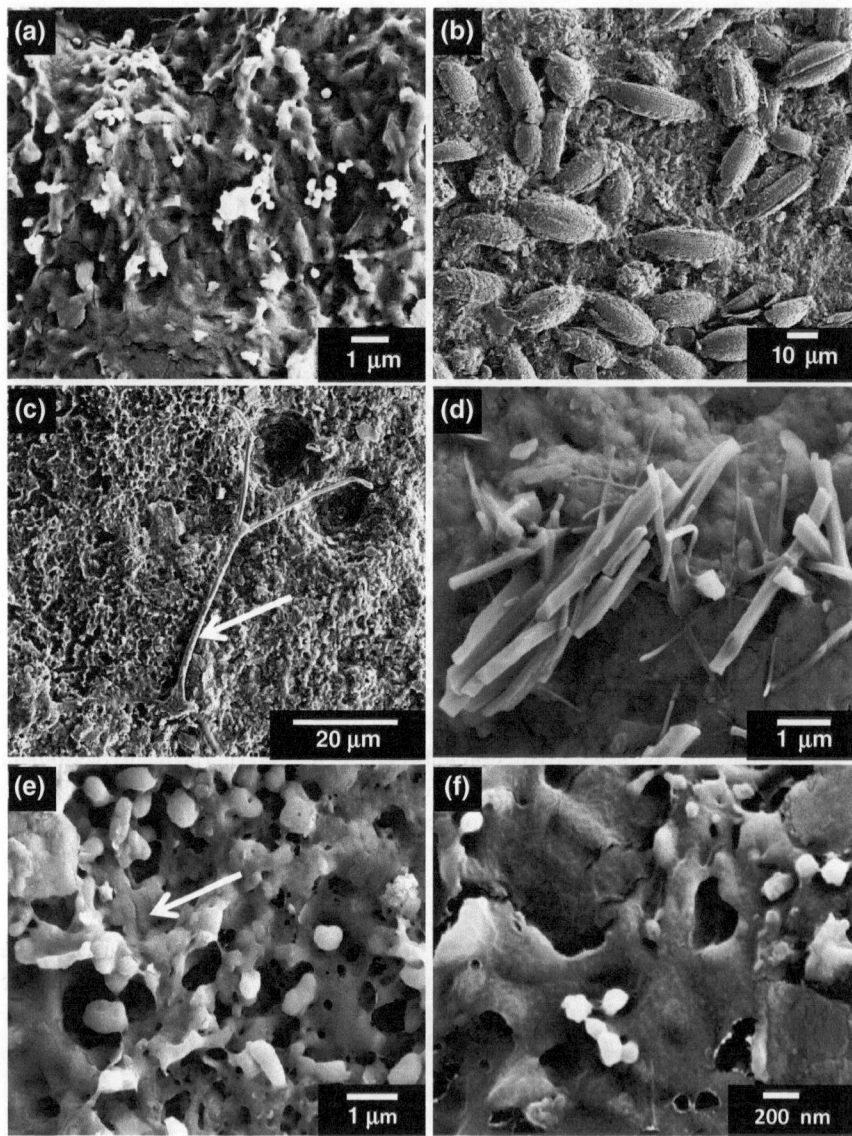

Fig. 2.8 SEM micrographs showing the colonization of various microorganisms on the surface of (**a, b**) P(3HB); (**c, d**) P(3HB-*co*-5 mol% 3HV) and (**e, f**) P(3HB-*co*-5 mol% 3HHx) films at fourth week of incubation in mangrove sediment. Bacteria with coccal morphology seen attached to the PHA surface in **a** and **f**. Pennate diatoms were seen dividing on the film surface in **b** and **d**. In **c**, *arrow* indicates the growth of fungal hyphae on the film surface whereas hemispherical cavities caused by bacterial degradation are indicated by *arrows* in **e**

while phospholipases act on phospholipid to release fatty acids and lipophilic molecules. Lipases are hydrolases, which catalyze the hydrolysis of carboxyl ester bonds of triglycerides (triacylglycerol) to release the fatty acids, monoglycerides

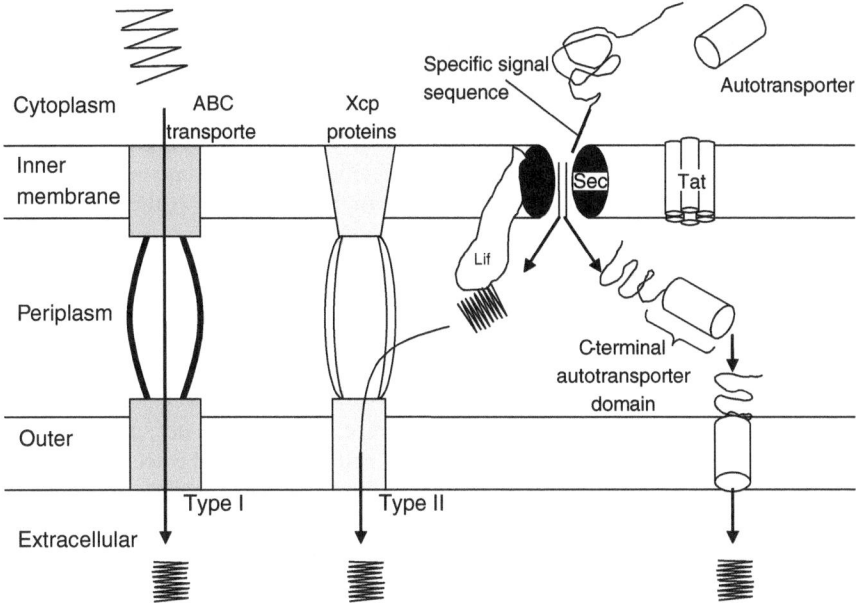

Fig. 2.9 Lipolytic enzyme translocation pathways used by bacteria (Jaeger and Eggert 2002; Rosenau and Jaeger 2000)

(monoacylglycerols), diglycerides (diacylglycerols), and glycerols. Lipases are also known as carboxylesterase which catalyze the hydrolysis and synthesis of long-chain glycerides. Although the hydrolysis activity of a lipase does not need any cofactor, the enzyme conformation has to be activated by interaction with the interfaces of substrate and aqueous solution (Grochulski et al. 1993; Joseph et al. 2008). Lipases have become important bio-catalyst in industries of food, detergent, chemical, and biomedical due to the wide range of substrate specificities and reactions (Jaeger et al. 1999).

Lipolytic enzymes of a variety of bacteria, including *Pseudomonas* sp. (Dharmsthiti and Kuhasuntisuk 1998; Ogino et al. 2007; Rosenau and Jaeger 2000), *Burkholderia* sp (Gupta et al. 2005a; Lee and Parkin 2003) and *Bacillus* sp. (Karpushova et al. 2005; Shi et al. 2010) have been investigated. Recently, the expression of lipase and esterase from *Ralstonia* sp. or *Cupriavidus* sp. in *E. coli* had also been studied (Quyen et al. 2005; Quyen et al. 2007). In a study, the lipase of *Pseudomonas aeruginosa* was reported to act like PHA depolymerase and exhibited hydrolysis activities on linear PHA such as poly(6-hydroxyhexanoate) or poly(4-hydroxybutyrate) (Jaeger et al. 1995). On the other hand, although PHA depolymerases have catalytic centers that are structurally similar to that of lipases, they do not show any lipase activity (Jaeger et al. 1995; Kumar et al. 2000).

For the oil-utilizing bacteria, extracellular lipolytic enzymes are secreted to the medium. As proteins, lipolytic enzymes need to be translocated across the membrane of bacteria. There are four pathways for lipolytic enzyme translocation in bacteria (Jaeger and Eggert 2002; Rosenau and Jaeger 2000) (Fig. 2.9).

The lipolytic enzymes secreted by Gram-positive bacteria cross only one layer of membrane using two pathways mediated by Sec and Tat multisubunit protein complex. The specific signal protein sequences which are contained in the lipolytic enzyme decide the pathway to be used. A lipolytic enzyme that possesses N-terminal signal sequence can be the substrate for the Sec protein exporter, while the lipolytic enzyme containing the specific Tat signal sequence can cross the Tat pathway (Fekkes and Driessen 1999). Both Sec and Tat pathways are also found in Gram-negative bacteria, but additional pathways are needed to cross the second membrane (outer membrane). After a lipolytic enzyme sets cross the first membrane and enters the periplasm between the bilayer membrane, the signal sequence cleaves from the lipolytic enzyme. The lipolytic enzyme is folded into an active conformation catalyzed by a cognate foldase (Lif) and transported by a complex protein (Xcp proteins) to the extracellular medium (Type II secretion pathway) (Jaeger et al. 1994). The esterase (EstA) of *P. aeruginosa* which possesses N-terminal domain to attach to the cell surface can be secreted via the outer membrane mediated by its C-terminal autotransporter domain (Wilhelm et al. 1999; Wilhelm et al. 2007). Other lipolytic enzymes which lack the N-terminal signal sequence but possess a C-terminal targeting signal are transported by the exporters consisting of three mediated proteins (ABC-transporter) (Binet et al. 1997). This type I pathway can transport the lipolytic enzymes directly through the bilayer membrane without folding into active conformation or needing a second transporter.

Chapter 3
Plant Oils and Agricultural By-Products as Carbon Feedstock for PHA Production

Abstract Many types of fermentation feedstock have been studied for the production of polyhydroxyalkanoate (PHA). Several industrial-scale processes have been developed for PHA production from sugars. Sugars are attractive feedstock because of their abundant supply worldwide, market stability, and also because the metabolism of PHA from sugars is very well understood. Recently, plant oils have been gaining much interest as a potential feedstock for PHA production. Industrial-scale processes for the production of PHA from plant oils are currently being developed. This chapter looks at the challenges in using plant oils, especially palm oil as feedstock for PHA production.

Keywords Crude palm kernel oil (CPKO) • β-oxidation • Fatty acids Fermentation • Lipase • Malaysia • Palm oil • Plant oils

In the case of microbial fermentation, selection of cheap and sustainable starting material is a critical factor which determines the overall performance of the fermentation process as well as production cost of polyhydroxyalkanoate (PHA). Therefore, the best approach is to choose readily available carbon substrates (preferably bio-based) which could be obtained in large quantities. This will ensure a lower substrate cost thus reducing the overall production cost. Besides, it is also necessary to identify substrates that can support efficient microbial growth and high PHA yield. The most conventional and popular carbon feedstocks used for microbial growth and PHA accumulation are sugars such as glucose, fructose, and other saccharides. Nowadays, plant oils are being investigated as suitable alternatives because they are renewable and produce higher yields of polymer. Utilization of plant oils such as soybean oil, palm oil, and corn oil as carbon source developed with findings of several studies which revealed that overall production cost and energy consumption could be reduced by using plant oils for PHA production (Akiyama et al. 2003).

Soybean oil (Kahar et al. 2004) and hydrolyzed corn oil (Shang et al. 2008) have been investigated for PHA production. Majid et al. (1994) reported the first attempts to use palm olein as the sole carbon source to produce P(3HB). Saponified PKO was first reported as a suitable substrate for the production of mcl-PHA using *P. putida* by Tan et al. (1997). Recently, various types of palm

K. Sudesh, *Polyhydroxyalkanoates from Palm Oil: Biodegradable Plastics*, SpringerBriefs in Microbiology, DOI: 10.1007/978-3-642-33539-6_3, © The Author(s) 2013

oil products were further evaluated as potential carbon feedstock for PHA production (Loo et al. 2005b; Annuar et al. 2007; Lee et al. 2008; Bhubalan et al. 2008b, 2010a, b). Besides plant oils, agricultural by-products and food industrial wastes are also being considered as inexpensive carbon and nitrogen sources for PHA biosynthesis. Bioconversion of whey (Ahn et al. 2001; Koller et al. 2008), molasses (Solaiman et al. 2006; Kulpreecha et al. 2009), starchy wastewater (Yu 2001), triacylglycerols (Solaiman et al. 2002), waste glycerol (Koller et al. 2005; Cavalheiro et al. 2009), used cooking oil (Vidal-Mas et al. 2001; Kek et al. 2008c), and other substrates from food industrial wastes or industrial by-products (Castilho et al. 2009) to value-added PHA polymers have been investigated. This is done in an effort to generate value-added green material from waste materials. P(3HB-co-3HV) copolymer was produced by *Comamonas* sp. EB172 in fed-batch cultivation using organic acids derived from POME (Zakaria et al. 2009, 2010b). Synthesis of copolymer with 3HV monomer composition up to 21 mol% was demonstrated.

3.1 Why Plant Oils are Potential Carbon Feedstock for PHA?

Although many studies had been conducted to optimize the production of PHA from sugars, the cost of production is still relatively high when compared to PHA yield obtained. The yield coefficient of P(3HB) production from glucose and sucrose ranges within 0.30–0.40 g per gram of glucose or sucrose used (Akiyama et al. 2003). In other words, to produce one kg of P(3HB) 3 kg of sugar is needed. It had been predicted that oil-based carbon sources provide higher yield for both cell biomass and PHA production.

Higher concentration of acetyl-CoA could be obtained via β-oxidation of oils compared to glycolysis of sugar compounds. As a result, higher total cell biomass and PHA content could be obtained. Theoretically, the degradation of a molecule of lauric acid which contains 12 carbons gives rise to 6 acetyl-CoA, which potentially leads to the formation of three 3HB monomer units. The same equation is applicable to other fatty acids such as myristic acid, which yields 7 acetyl-CoA (three and a half 3HB monomers), palmitic acid which yields 8 acetyl-CoA (four 3HB monomers), and so forth. In the case of glucose, only two molecules of acetyl-CoA are produced via Entner-Doudoroff pathway, which potentially lead to the generation of one 3HB monomer.

With respect to the ability to produce high density cultures, Kahar et al. had shown that soybean oil is a desirable carbon feedstock. Up to 128 g/L of cell biomass and 74 wt% of P(3HB-co-3HHx) copolymer was produced from *C. necator* transformant harboring the PHA synthase gene of *A. caviae*. Similar experiment was also conducted by using wild-type *C. necator* H16, which produced 126 g/L of cell biomass and 76 wt% of P(3HB), respectively. In general, the yield of PHA from soybean oil was in the range of 0.72–0.76 g/g and the PHA productivity was approximately 1 g/L/h (Kahar et al. 2004). In a separate study, the productivity of

P(3HB-co-3HV) copolymer synthesized from a mixture of corn oil and propionic acid by *C. necator* was recorded at 1.76 g/L/h (Naylor and Wood 1999). In addition, Naylor and Woods also demonstrated that high P(3HB-co-3HV) yields ranging from 0.64 to 0.72 g could be obtained from 1 g of the carbon source (mixtures of rapeseed oil or corn oil with propionic acid) by using *C. necator* (Naylor and Wood 1999).

According to the stoichiometric equation proposed by Akiyama and coworkers, it was estimated that 1.38 g of P(3HB) could be produced from 1 g of linoleic acid (the main fatty acid composition in soybean oil). The estimated cost of P(3HB-co-3HHx) production from soybean oil and glucose was USD 3.50/kg and USD 3.80/kg, respectively (Akiyama et al. 2003). It should be noted that antibiotic was used to maintain the plasmid containing PHA synthase of *A. caviae*. In an attempt to eliminate antibiotic usage, studies were carried out to integrate the PHA synthase into the chromosome of the host (Mifune et al. 2008). To date, results on PHA production using plant oils and agricultural by-products seems to be promising as positive findings are constantly being reported by researchers around the globe. The only factor that may affect the potential usage of plant oils is its price.

3.2 Effect of Fatty Acids on Cell Growth and PHA Accumulation

Plant oils in the culture medium are hydrolyzed by extracellular lipases excreted by oil-utilizing strains. This results in the generation of monogylcerides, diglycerides, free fatty acids, and glycerol. The fatty acids are then transported into the cells freely in an undissociated form (non-ionic diffusion) (Salmond et al. 1984). In the cell cytoplasm, these fatty acids are readily converted into acyl-CoA esters via β-oxidation pathway to minimize inhibitory effects. It is noted that coordinated induction of important β-oxidation enzymes and fatty acid transport system is necessary for bacterial cells to grow on fatty acids. The ability of fatty acids to induce the enzymes is dependent on its chain length. In theory, fatty acids of carbon chain length of 14 or longer induce the necessary enzymes of fatty acid oxidation and readily support growth, whereas shorter carbon chain length fatty acids do not induce the enzymes of fatty acid oxidation (Iram and Cronan 2006).

It was found that, cell growth was inhibited when shorter chain length fatty acids such as caproic acid (6C) and caprylic acid (8C) were used as the sole carbon source (Iram and Cronan 2006). Earlier report by Kahar et al. suggested that *C. necator* favored fatty acids such as palmitic acid (C16:0), oleic acid (C18:1), and linoleic acid (C18:2). In contrast, linolenic acid (C18:3) was poorly consumed by the bacterium (Kahar et al. 2004). In addition, the residual fatty acids from soybean oil in culture medium was confirmed with thin layer chromatography (TLC) and it was observed that only certain fatty acids were utilized for cell growth and PHA accumulation. This might be due to the accumulation of linolenic acid in the culture medium which had limited the transfer of fatty acids into cells. This was supported by a finding from O'Leary whereby the inhibitory effects of unsaturated fatty acids on the growth of bacteria increases with higher number of double bonds in the molecule (O'Leary 1962).

Crude palm kernel oil (CPKO) which is derived from the kernel of the oil palm fruit consists of large amounts of saturated fatty acids such as lauric acid (C12:0) [~48 %], myristic acid (C14:0) [~16 %], and palmitic acid (C16:0) [~8 %] (Loo et al. 2005). However, the concentration of unsaturated fatty acids such as linoleic (C18:2) [~2 %] is very low. It contrast, palm oil products derived from mesocarp such as crude palm oil (CPO) and palm olein (PO) are mainly composed of C16:0 and contain more unsaturated fatty acids such as oleic acid (C18:1), (C18:2) and trace quantities of linolenic acid (C18:3) (Loo et al. 2005). Compared to palm oil products, soybean oil is rich in unsaturated fatty acids with C18:2 (54 %), C18:1 (22 %), and C18:3 (8 %) as the major constituents while 10 % of the distributions are contributed from saturated fatty acids (Kahar et al. 2004; Loo et al. 2005). Nevertheless, soybean oil has proven to be a good carbon source for high cell density cultures. Since, CPKO contains lesser unsaturated fatty acids, it could become potential carbon feedstock for high cell density PHA production.

3.3 Challenges in Using Plant Oils as Carbon Feedstock for PHA Production

There are few challenges to be addressed before plant oils could be used for large-scale PHA production. First, strains capable of growing and accumulating PHA using triglycerides have to be identified. This is because, even if a bacterium possesses high cell growth rate on plant oils or fatty acids, their ability to accumulate PHA is sometimes low (Tsuge 2002). In addition, plant oils that are rich in saturated fatty acids such as palm stearin usually occur in solid form at 30 °C. Difficulties arise when these oils are employed as carbon feedstocks for bacteria which grow best at 30 °C. In solid form, breakdown of the triglycerides into free fatty acids by lipase might be reduced thus affecting the uptake of fatty acids into bacterial cell. Hence, the cell growth and PHA accumulation are very low. Nevertheless, this problem could be possibly overcome by incubating these feedstocks at temperature above 30 °C while feeding is done. Although PHA homopolymer or copolymer possessing interesting properties could be produced from plant oils, only certain microorganisms could directly use plant oils as the carbon feedstock. This is mainly due to the inability of some microorganisms to secrete extracellular lipase. Therefore, an additional saponification step would be required to supply free fatty acids to the bacterial cells (Tan et al. 1997). This strategy might not be efficient from the economic point of view.

When plant oils are hydrolyzed by extracellular lipase, various fatty acids with different carbon chain length and glycerol would be released into the culture medium. Fatty acid such as lauric acid is known to induce foaming which will require the addition of anti-foam (Chen et al. 2001). The presence of anti-foam may not be favorable for cell growth due to possible toxicity effect. In addition, it could affect the cost of production when the fermentation is conducted in industrial scale (Chen et al. 2001). Anti-foam may also interfere with the downstream processes of extraction and purification of the PHA granules. The breakdown of oils by lipase also produces glycerol molecules. The presence of glycerol in the culture medium is of some concern, as it is known that glycerol affects the molecular weights of PHA by acting as chain transfer agents in the chain termination step of polymerization (Madden et al. 1999). However, in some cases, the M_w of polymers synthesized using oily substrates are high and comparable to those synthesized using other carbon substrates (Lee et al. 2008; Bhubalan et al. 2008). Availability of local supply of plant oil is also essential in order to ensure a cheap and sustainable PHA production. Plant oils such as soybean oil and palm oil are available in bulk quantities in several countries such as The United States of America, Malaysia, and Indonesia. Application of these oils as starting material for large-scale PHA production will be feasible as PHA could be marketed at reasonable price and this will boost the PHA industries.

3.4 Palm Oil: A Potential Renewable Feedstock for PHA Production

Plant oils are derived from oil-bearing crops, such as soybean, rapeseed, palm, sunflower, and corn. Interest abounded in the use of plant oils in the bio-based product industry, more specifically, for PHA production since the beginning of the last decade. Prevailing research have unveiled a number of potential oil-based candidates as carbon feedstock, namely soybean oil, hydrolyzed corn oil, and various palm products (Majid et al. 1994; Tan et al. 1997; Fukui and Doi 1998; Kahar et al. 2004; Loo et al. 2005; Bhubalan et al. 2008; Lee et al. 2008; Shang et al. 2008; Kek et al. 2010). Plant oils have an edge over other conventional and well-known carbon feedstocks such as sugars in terms of price competitiveness and the ability to produce higher yields of PHA (Akiyama et al. 2003).

Since the 1950s, soybean oil has been the leading plant oil in production and in use worldwide, with rapeseed oil close behind. Spurred by income and population growth in developing countries as well as rapidly expanding food processing industries in Asia, an astounding expansion of oil palm plantation in Malaysia from the total planted area of 54,000 ha in 1960 to 4.69 million ha in 2009 materialized (MPOB 2009), outpacing that of most other agricultural products (Basiron 2007). Today, Malaysia stands as one of the largest contributor of palm oil in the world, surpassing Nigeria as the main producer since 1971 (Yusoff 2006). Malaysia alone exported a record breaking amount of 15.87 million tonnes of palm oil to the international palm oil market in 2009. The total exports of oil palm products consisting of palm oil, palm kernel cake, oleochemicals, and finished products amounted to 22.4 million tonnes in 2009 (MPOB 2009).

Palm oil is obtained from the fruit of a palm tree, *Elaeis guineensis*, native to tropical West Africa and cultivated in Africa, Indonesia, Malaysia, and tropical America. The palm fruit grows in large bunches weighing approximately 10–20 kg which could be made up of 1,000–2,000 individual fruitlet, consisting of a fleshy mesocarp outer later and hard whitish kernel. Being a perennial tree, this crop bears fruit throughout the year. In addition, each oil palm tree continues producing fruit economically for up to 25 years (Sambanthamurthi et al. 2000b). This ensures a constant and stable supply of palm oil compared to other temperate oilseed crops.

Oil palm fruit is unique compared to other oil-producing crops as different types of oils could be extracted from the mesocarp and kernel of the oil fruit, respectively. Crude palm oil (CPO) is the primary product obtained from the mesocarp while crude palm kernel oil (CPKO) is derived from the kernel. Further chemical and physical refining results in various palm oil products such as CPO, palm olein (PO), palm stearin (PS), refined, bleached, and deodorized (RBD) palm oil, kernel olein, and kernel stearin as well as by-products. Furthermore, the by-products that have been produced can also be manufactured into useful products. The by-products include palm kernel acid oil (PKAO), palm acid oil (PAO), and palm fatty acid distillate (PFAD) (Fig. 3.1). The palm oil by-products are generated from the refining processes during the removal of free fatty acids from CPO which are detrimental to the oil properties. More specifically, PAO and PKAO are the by-products from chemical refining process, while PFAD is generated from physical refining process. Most of the oil exported by Malaysia is in the form of RBD palm oil. These ranges of products suit a variety of manufacturing needs in the forms that are ready-to-use and require no further processing. The oils derived from oil palm fruit could be utilized for edible as well as non-edible applications. The fatty acid compositions of various palm oil products are shown in Table 3.1.

On account of feedstock availability, Malaysia's position as a major global producer of palm oil would firmly support the supply of feedstock for the PHA industry, at the same time creating a significant elevation in the overall value of the palm oil industry. Some general information about oil palm and the palm oil industry in Malaysia can be found in Table 3.2. There are, however, issues requiring serious attention such as deforestation, waste disposals from palm oil mill, and energy expenditure when PHA is to be produced in large scale. Also, some view the bioconversion of food crops to PHA as a practice that carries a high risk of dwindling world's food supply. It should be noted here that palm oil provides nearly 30 % of the world's edible vegetable oil (Carter et al. 2007), with a production volume of 43.12 million tonnes in year 2009 (MPOB 2009). Therefore, a more sustainable way of producing palm oil should be practised through improved processes and better management practices.

The global demand for bioplastics was estimated at 0.36 million tonnes, which is equivalent to 0.2 % of the annual petrochemical plastic production (Thompson et al. 2009). PHA accounted for about 10 % of the bioplastic market which is currently dominated by poly(lactic acid) and other starch-based biopolymers (Barker et al. 2009). Based on the statistics, in order to fulfill the PHA market demand

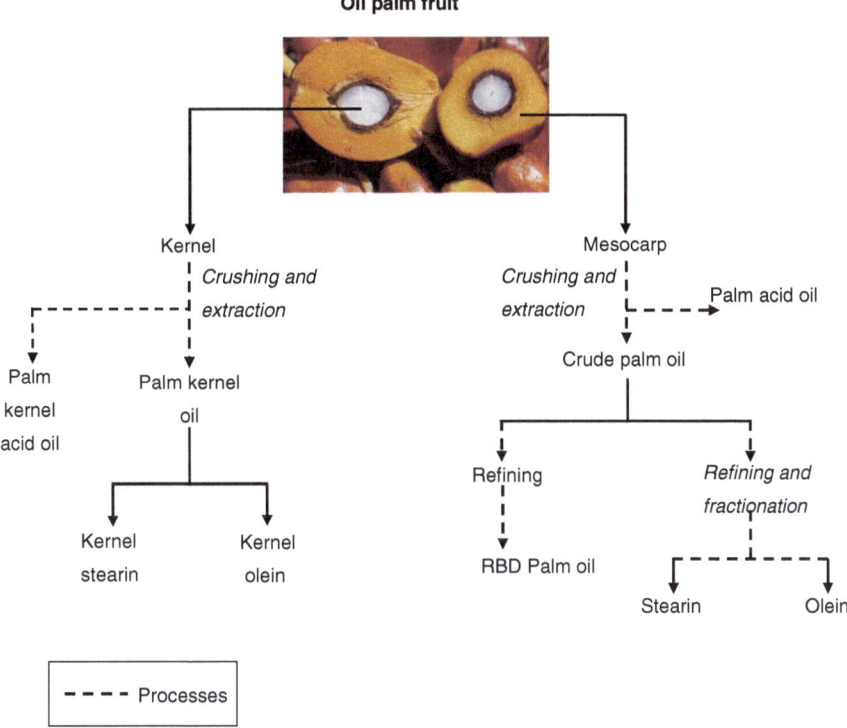

Fig. 3.1 Cross-section of a typical oil palm fruitlet and processes involved in the extraction of various oil fractions in a typical palm oil milling industry (Maycock 1992)

solely by using CPKO-derived PHA, approximately 53,000 tonnes of CPKO (~2.8 % of Malaysia's total CPKO production) is required as carbon feedstock for microbial fermentation. In other words, the production of 52,000 tonnes of PHA per annum would involve a total of 111,520 hectares of oil palm plantation; approximately 2.6 % of total oil palm planted area in Malaysia. With the average price of CPKO at US$ 0.76 per kg (MPOB 2009), the estimated carbon substrate cost per kg of PHA produced is US$ 0.78. By the year 2020, global demand for bioplastics is estimated to reach 1.5 % of total world plastic production, which would account for approximately 3.45 million tonnes per annum (Shen et al. 2010).

3.5 Characteristics of Palm Oil

Oil palm fruit is unique compared to other oil-producing crops as two different types of oils could be extracted from the mesocarp and kernel of the oil fruit, respectively (Fig. 3.1). CPO is the primary product obtained from the flesh or the mesocarp

Table 3.1 Fatty acid compositions and oil properties of various palm oil products

Fatty acid	CPO	PAO	CPKO	PKAO	PFAD	PS	PO	WFPO
Saturated								
Caproic (6:0)	–	–	–	–	–	–	–	–
Caprylic (8:0)	–	–	3.9	2.0	–	–	–	–
Capric (10:0)	–	–	3.5	2.5	–	–	–	–
Lauric (12:0)	0.1	0.8	48.5	44.1	0.2	0.2	0.2	0.5
Myristic (14:0)	0.9	1.1	16.2	17.8	1.2	1.1	1.0	0.9
Palmitic (16:0)	43.8	44.7	7.5	10.8	47.2	58.5	35.8	37.1
Stearic (18:0)	4.0	3.7	2.6	3.1	4.5	5.0	4.1	4.8
Arachidic (20:0)	–	–	–	–	–	–	0.4	–
Unsaturated								
Palmitoleic (16:1)	–	–	–	–	–	–	0.1	–
Oleic (18:1)	42.1	40.3	15.7	17.3	36.7	5.0	43.8	42.6
Linoleic (18:2)	8.9	9.4	2.1	2.3	9.7	28.0	14.3	10.5
Linolenic (18:3)	0.2	–	–	–	0.47	7.1	0.2	–
Oil properties								
Iodine value	51–55	41.8–64.4	16.2–19.2	–		51.2–57.4	21.6–49.4	56.1–60.6	55.9
FFA (max.%)	5	72.8	5	–	83.3	0.08–0.10	0.45	4.5
Moisture, impurities (max.%)	0.25	0.28	0.5	–	0.08	0.01	0.02	–
MP (°C)	24	–	26–28	–	–	44.5–56.2	19.4–23.5	–

CPO crude palm oil, *PAO* palm acid oil, *CPKO* crude palm kernel oil, *PKAO* palm kernel acid oil, *PFAD* palm fatty acid distillate, *PS* palm stearin, *PO* palm olein, *WFPO* waste frying palm oil (*Source* Unitata Ltd.; Gunstone 2002; Shahidi 2005; Lertsathapornsuk et al. 2008)

while CPKO is derived from the kernel (Basiron 2007). The oils derived from oil palm fruit could be utilized for edible as well as non-edible applications. A mature oil palm fruit is harvested through cutting the bunch from tree and allowing it to fall onto the ground. In the conventional milling process, the fresh fruit bunches are threshed to remove the mesocarp layer of the fruits from the oil palm bunches. The mesocarp is then sterilized, digested, and pressed either through wet or dry method to extract its oil content, which is known as CPO. On the other hand, the nuts are separated from the fiber through heat, dried, and crushed to obtain the CPKO.

CPO and CPKO mainly consist of glycerides and small portions of non-glyceride components. Non-glycerides include free fatty acids, phospholipids, tocopherols or tocotrienols, oxidation products, and trace metals. In order to make this oil edible, the non-glycerides components have to be removed through chemical processes which include refining, bleaching and deodorizing. Fractionation of CPO and CPKO produces liquid olein and a solid stearin component. Naturally, palm oil is considered as stabilized oil due to its chemical composition. Hence, it can be used in most food applications without hydrogenation. This helps to reduce production cost as much as 30 %. Palm oil is found in a variety of forms such as CPO, palm olein (PO), palm stearin (PS), refined, bleached and deodorized (RBD)

Table 3.2 Information on oil palm and the palm oil industry in Malaysia

			References
Oil palm	Total planted area (million hectares)	4.69	(MPOB 2009)
	Plantation density (palms/hectare)	148	(Basiron et al. 2004)
	Economical life span I (years)	25	(Sambanthamurthi et al. 2000b)
Fresh fruit bunch (FFB)	Weight (kg)	10–20	(Corley and Gray 1976; Sambanthamurthi et al. 2000b)
	Number of fruits/FFB	1,500–2,000	
	Average number of FFB/palm	[a]10	
Palm oil	Production volume (million tonnes/annum)	19.58	(MPOB 2007a, b, 2009)
	Average yield (million tonnes/ hectare/annum)	[b]15.82 [c]1.91	
	Average market price (US$/tonne)	[b]727.43 [c]758.87	
	Edible use fraction (%)	74	(USDA 2005)
	Amount of by-product/wastes generated by palm oil mills (million tonnes/annum)		(MPOB 2007c; Hassan et al. 2006)
	(a) Empty fruit bunch	15.8	
	(b) Fronds	12.9	
	(c) Mesocarp fiber	9.6	
	(d)Trunk	8.2	
	(e) Shell	4.7	
	(f) Palm oil mill effluent	30	
	Amount of water required by palm oil mill (tonne/tonne of CPO)	5.0–7.5	(Ahmad et al. 2003)
	Cost of production (US$/tonne of palm products)		(SimeDarby 2009)
	(a) Estate cost	256.36	
	(b) Mill cost	62.81	

[a]Value shown represents the quantity present at any one time
[b]Crude palm oil
[c]Crude palm kernel oil

palm oil, fractionated palm olein and palm mid-fraction. Most of the oil exported by Malaysia is in RBD palm oil and RBD palm olein form. This range of products is available to suit a variety of manufacturing needs in the forms that are ready-to-use and require no further processing.

The mesocarp comprises about 70–80 % by weight of the fruit and about 45–50 % of this mesocarp is oil. The rest of the fruit comprises the shell, kernel, moisture, and other non-fatty fiber. CPO derived from mesocarp is similar in

composition like all other natural fats and oils which mainly comprises of triglycerides, mono-, and diglycerides. Free fatty acids, moisture, dirt, and minor components of non-oil fatty matter are collectively referred to as unsaponifiable matter.
The property of a triglyceride will depend on the different fatty acids that combine
to form it. The fatty acids themselves are different depending on their chain length
and degree of saturation. Short-chain fatty acids are of lower melting point and are
more soluble in water. Whereas, the longer chain fatty acids have higher melting
points and are less soluble in water. The melting point is also dependent on degree
of unsaturation. Unsaturated acids will have a lower melting point compared to saturated fatty acids of similar chain length. The two most predominant fatty acids in
CPO are C16:0 (saturated) palmitic acid and C18:1 (unsaturated) oleic acid. CPKO
is considered as a secondary product of the palm oil industry and is obtained from
the endosperm (kernel) of palm fruit. It constitutes about 45 % of the overall kernel
weight. On a wet basis, the kernel contains about 45–50 % of oil. The CPKO and
CPO differ greatly in their characteristics and properties even though they originate
from the same fruit. CPKO is rich in C12:0 (48.3 %) and the other major fatty acid
components are C14:0 (15.6 %) and C18:1 (15.1 %) (Goh 1993). The lower composition of unsaturated fatty acid gives CPKO a solid consistency at cool ambient temperatures, but nevertheless it melts at temperatures above 30 °C (Nonato et al. 2001).

3.6   The Palm Oil Agro-Industry in Malaysia

The production of plant oils (i.e., palm oil, soybean oil, rapeseed oil, sunflower oil,
corn oil) reached a total of 129 million tonnes in 2007 as compared to 125 million
tonnes and 117 million tonnes in 2006 and 2005, respectively. Palm oil products
are currently the most productive plant oils in the global market with the production volume of 38.2 million tonnes in 2007. This is followed closely by soybean
oil which amounts to 37.4 million tonnes and rapeseed oil at 18.5 million tonnes.
The oil palm cultivation in Malaysia has a history of approximately 90 years
(Jaqoc 1952). The initial growth of the industry was very slow and only expanded
at incredible speed in the late 1950s following the launch of major government
policies which were targeted to diversify the country's agricultural production and
to raise the socio-economic status of the increasing population.

Oil palm was then earmarked as one of the most potential oil yielding crop
which subsequently caused a rapid expansion of the oil palm plantation at the
expense of other crops such as rubber and cocoa or through planting on idle lands
(Basiron 2007). Being a perennial tree, this crop could be harvested all year round.
This ensures a constant and stable supply of palm oil compared to other annual
crops. The commercial oil palm trees planted in countries such as Malaysia,
Indonesia, and other South East Asian regions are believed to have originated
from West Africa (Hartley 1989; RMRDC 2004). In Malaysia, this crop was first
introduced as an ornamental plant in 1870 and the first small commercial oil palm
estate was set up in 1917 (Jaqoc 1952).

Chapter 4
Is Palm Oil Produced in a Sustainable Manner?

Abstract The most important aspect of any feedstock for industrial-scale production of polyhydroxyalkanoate (PHA) is market stability. One would expect the feedstock to be sustainable in terms of supply, cost and quality. In addition, recently, there is also growing concerns over the use of food-grade feedstock for making nonedible products such as fuel and material. Therefore, the selection of a feedstock for PHA production must take into consideration the effect on global food supply. This chapter presents the current scenario of the palm oil industry along with issues such as land management and conservation of biodiversity. In order to ensure the sustainability of PHA production from palm oil, several strategies are proposed.

Keywords Biodiversity • Malaysia • Palm oil • Polyhydroxyalkanoate (PHA) • Sustainability • Spent cooking oil • Palm oil by-products

Is palm oil currently or ever produced in a sustainable manner? This is often the question of many environmentalists and the concern of public. Most people are convinced that rapid expansion of oil palm plantation over the years has left irreversible impacts on the environment including aspects such as land deforestation, extinction of several indigenous species and peat land destruction (Tan et al. 2009). Nevertheless, stakeholders and the parties involved intimately in oil palm plantation and palm oil production are convinced that these criticisms are baseless and the real scenario has been exaggerated.

The simplest definition of sustainability refers to the ability of mankind to reach their needs at present and at the same time not compromising the needs and capacity for future generations. Well-established oil palm plantations seem to meet these standards, with an exclusion that these plantations were done at the expense of deforestation. This act is deemed unsustainable, because destruction of pristine forests has permanent effect on the global ecosystem and future sustainability, thus, leading to the conclusion that deforestation is intolerable. Nevertheless, Robinson reviewed that in order for this present generation to meet their needs, agricultural developments are inevitable; and naturally for these activities to take place, land is needed (Robinson 2004).

K. Sudesh, *Polyhydroxyalkanoates from Palm Oil: Biodegradable Plastics*,
SpringerBriefs in Microbiology, DOI: 10.1007/978-3-642-33539-6_4,
© The Author(s) 2013

From being a minor crop in the 1960s, palm oil production had increased steeply and overtook soybean oil as the most produced and traded plant oil in the global plant oils market in the year 2005. In Malaysia alone, the annual production of CPO has seen an increase from 1.3 to 15.9 million tons in the past 30 years (OW 2007). This exponential increase in palm oil production is in tandem with growing human population and their consumption of plant oils for various applications. Looking back at the history of mankind, it can be seen that the increase in the consumption of fats and oils is often associated with economic boom and increasing prosperity. During the height of economic boom in the 1990s, people with higher income had changed their nonfatty dietary preferences to high-oil diets (Murphy 2007). This was also the time when the demand for plant oils far exceeded the other food products due to increased consumption. Therefore, the demand for palm oil is expected to ever increase with the rising human population and stronger purchasing power as compared to 40 years back. Hence, the expansion of oil palm plantation and palm oil-based industries are inevitable. It is not surprising to learn that palm oil sectors had been the prime target for criticisms by environmentalists for imposing lasting damage on the environment. In line with this, it is imperative to discuss several issues surrounding the sustainability concerns. The main issues surrounding the 'un-sustainability' of oil palm is driven by two major concerns; land management and biodiversity.

4.1 Land Management and Conservation of Biodiversity

There is no denying that conversion of forests to any mankind activities to certain extent causes detrimental damage on the environment. Going back to issues of agricultural activities, the expansion of oil palm plantations had been at the expense of pristine forests. Several reports had also pointed out that palm oil producers preferred to clear pristine forests for extra income from selling the timbers rather than planting on abandoned lands (WWF 2005). In other words, no matter how sustainable or environmental-friendly techniques have been used in the oil palm plantation, it is still deemed unsustainable if the activities were carried out at the expense of forest areas. Statistics had revealed that a mere 38,000 ha of lands were used for oil palm plantations in 1950 while more than 4.17 million ha of lands were put aside for the same activity in 2006 (MPOB 2008). This revelation is disconcerting as it indicated that the criticisms hurled by the environmentalists were somehow true.

However, some of these observations are contradictory to the reports by Malaysian Palm Oil Council (MPOC 2007). Oil palm is planted on 4.17 million ha, which is less than 10 % of the total land area of Malaysia (32.85 million ha as at the end of 2006). Furthermore, out of the total land areas in Malaysia, a mere 19 % or 5.4 million ha is used for agricultural purposes. In comparison, an astounding 70 % of lands in Malaysia are still available and unexplored. The country has at least 25 million ha of its total land area, under forest and in national parks, wildlife sanctuaries and nature reserves (MPOC 2007). The Malaysian Government had also made it clear that the gigantic expansion of oil palm plantations does not infringe on forest areas. The increase in oil palm plantation areas was done either through the planting on idle land or conversion from other lesser 'economical value' crops such as rubber, coconut, and cocoa. Through this strategy, more than 1 million ha of land (which was originally used for other crops) had been replanted with oil palm (Basiron 2007). Therefore, the 'real' expansion of oil palm plantation on new lands from 1990 to 2005 was less than 1 million ha, contrary to otherwise claimed. The comparison between oil palm crops and other major crops should also be made on the use of forest lands. In terms of productivity, palm oil is the most efficient oil-bearing crop in the world. The production of 16 million tons of oil in Malaysia was achieved with only 4.17 million ha of planted area, which represents less than 2 % of the total planted area of the world's oilseeds. Soybean oil, the second largest traded oil in the global market (33.6 million tons) required 92.5 million ha of land to produce similar output (Basiron 2007).

Malaysia and Indonesia, located at the Southeast Asia region, are currently the two major players in the palm oil market. Together, both countries account to more than 80 % of the global palm oil production. It is also a coincidence that both countries have been identified as the world's biodiversity hotspots, which harbor high numbers of indigenous and endemic species, which mostly inhabit the forests (Sodhi et al. 2004; Peh et al. 2006). Therefore, the excessive expansion of forest areas in Southeast Asia for oil palm activities seems the greatest threat to biodiversity (Noda et al. 2005; Donald 2004).

There is a need for equilibrium between agricultural development, the protection of rainforests, and natural habitats. In the past, the palm oil industries had been established on forest lands without proper consideration on its impacts on natural habitats and the diversity of flora and fauna. To safeguard the biodiversity in countries like Malaysia, ideally the reliance on the new forest areas for oil palm plantation should be reduced. However, if and when new plantations are urgently required, proper mitigation tools to assess the environment, biodiversity, and social effects should be established before such decision making (Fitzherbert et al. 2008). In addition to that, habitats rich with species diversity and of high conservation values should be identified as soon as possible and should be protected from future mankind activities (Koh and Wilcove 2008). Supplementary data on the land use have to be collected and analyzed at national level to recognize the effects of future conversions of forest lands to oil palm and other agriculture uses (Scharlemann and Laurance 2008; Reijnders and Huijbregts 2008). Apart from that, secondary forest should also be given the same consideration as pristine forests and protected from further exploitation. Ideally, the future expansion of oil palm should be confined to existing crops land and other degraded habitats.

It was also suggested that the biodiversity of existing oil palm plantations can be improved by sustainable agricultural practices through the concept of 'half-plantation and half-forest'. As stated by Koh, the increase of old-growth forests (from 0 to 23 %) and natural forests surrounding the oil palm plantations resulted in the increase of butterfly biodiversity by 3.7 and 12.9 species, respectively (Koh 2008). These data indicated that natural forests and old-forests surrounding the oil palm plantations are important for enriching the biodiversity of bird and butterfly species within the oil palm plantation. This idea is not new; in fact, human-made park adjoining to forests in Singapore had been shown to harbor higher number of butterfly species compared to the ones which lack the adjacent forests (Koh 2008).

4.2 Concern Over the Conversion of Plant Oils to Consumer Products

Plant oils such as soybean oil and palm oil are currently being investigated as potential carbon feedstocks for PHA production. However, these two types of oils are the most commonly used triglycerides in food industries. Various types of edible materials are derived from palm oil alone. The growing human population has created larger demands for edible oils as the starting material for preparing food products. A shortage in raw material is inevitable when soybean oil or palm oil is converted in excess or in a nonsustainable manner to PHA. In addition, the price of food products is expected to skyrocket and this in turn might create serious inflation to countries which rely heavily on imported food products (Berck and Bigman 1993). Therefore, it is important to manage and utilize surplus material such as palm oil in a judicious manner.

One example of conversion of food material to consumer product is biodiesel. Conversion of palm oil into biodiesel has been carried out in various countries in Europe as well as in Malaysia, a move motivated mainly by the increasing price of petroleum and to a certain extent to mitigate global warming by reducing CO_2 emission. Production of biodiesel in Malaysia increased twofolds from 47,986 tons in 2006 to 89,132 tons in 2007, which could only substitute a mere 1 % of the demand for diesel (Lam et al. 2009). The demand for biodiesel in Europe alone was expected to reach more than 10 million tons by 2010 (Lam et al. 2009). Nevertheless, by using biodiesel, exploitation of natural resources could be controlled and sustained for the future generation.

It is expected that there will be a shortage of substrate when both PHA and biodiesel are in active production. This is because, at in the current state, palm oil industry in Malaysia is still not ready to sustain the production of both materials in large scale. The world demand for edible triglycerides is expected to maintain at a minimal growth rate of 3 % per year. At the same time, palm oil industry in Malaysia is also expected to grow 3 % by year 2009 and subsequently to 10 % in future as more palm oil trees mature. Therefore, some time is needed before Malaysia is able to produce a larger amount of excess palm oil. Recently, an increase of excess palm oil was recorded from 1.5 million tons in 2005 to 1.7 million tons in 2007 (Lam et al. 2009). It is expected that the excess supply of palm oil will increase, since Malaysia is actively engineering more potential and effective oil palm seed to further improve the oil productivity. With cloned oil palm seed, the average productivity is expected to reach 10.6 tons of oil per ha per annum and it is 20–25 % higher than the yield from conventional seedling (Lam et al. 2009). At the same time, Indonesia is also aggressively developing its oil palm plantation and palm oil-based agro-industries since year 2000. Hence, bulk supply of palm oil is expected in several years to come. This will ensure uninterrupted supply of palm oil for food products and for the development of other commodity materials such as PHA and biodiesel. However, it should be noted that even if all the world's vegetable oils were to be converted into biodiesel, this will only substitute a small percentage of the world's demand for fuel.

4.3 Integrated System for Palm Oil and PHA Production

Successful large-scale production of PHA is largely determined by the availability and constant supply of cheap fermentative substrates. At the same time, the overall cost involved in the production of this biodegradable polymeric material needs to be controlled and reduced in order to penetrate and compete in the world's commodity market. Palm oil has been identified as suitable carbon feedstock and potential strains capable of utilizing this raw material have been discussed in the above sections. However, waste disposals from palm oil mill and the amount of energy needed for PHA production are other issues that require equivalent attention when PHA is produced in large scale.

To date, the integrated P(3HB) production system in sugar cane mills in Brazil had been appraised as the most appropriate, cheapest, and efficient for the production of PHA (Nonato et al. 2001). To reduce the reliance on the raw petroleum energy, the solid wastes from sugar agro-industry were used to supply additional energy for the industries as well as the PHA production system. This strategy also ensures that all the resources are completely utilized without wastages. Palm oil producing countries such as Malaysia also encourages the implementation of this concept into the palm oil mill either for PHA production system or for the palm oil processing. Approximately, 40 % of biomass such as oil palm fibers, empty fruit bunches, and kernel shells are generated from the palm oil mill. The average amount of whole biomass produced by a palm tree is approximately 231.5 kg per year (Mohamad et al. 2005). Meanwhile, the quantity of biomass wastes accounts for 51.2 metric tons per ha annually. These wastes contribute to US\$ 1.8 billion worth of energy annually (Jaafar et al. 2003). Biomass is potentially important for electricity or heat generation since they contain large amount of energy.

Besides that it is interesting to note that the CO_2 emission from coal or oil-based electrical power plants is much higher compared to that generated from biomass to generate the same amount of current supply. For every kWh, approximately 1,100 g of CO_2 could be produced from coal or oil. However, this figure could decrease almost 63 folds to 16 g when biomass is used instead. Interestingly, in most developing countries, biomass contributes to more than 40 % of the primary energy (Ryu et al. 1999). Biomass is a suitable raw material that can be utilized for the generation of power supply. This could ensure an environmental-friendly manner to obtain energy for the production of PHA. Ideally, the PHA industry should be established nearby the oil palm plantation or the palm oil mill. Cost for transportation of the palm oil products could be maintained at the most minimal level and oil palm biomass could be used for cleaner energy generation which is needed to run PHA production systems.

4.4   Is the Palm Oil Supply in Malaysia Sufficient for Continuous Large-Scale PHA Production in the Future?

The exploration of inexpensive agricultural products as fermentative substrates for PHA production in larger scale is useful to expedite the commercialization of bio-based plastics. The industrial production of PHA will become more profitable when carried out at locations where a constant supply of good fermentation substrate such as palm oil is readily available. Malaysia is a strategic location to develop palm oil-based PHA industries due to the abundant supply of palm oil products. It should be noted that the option of producing PHA from palm oil in other countries such as Brazil and United States might not be an attractive option due to the limited supply of this resource as most of it is imported. Instead, sugarcane and soybean oil might be the more rationale options in Brazil and United States, respectively. Recently, the non-edible jatropha oil was also studied as the carbon feedstock for PHA production (Ng et al. 2010). Jatropha oil is abundantly found in India. PHA could be produced in a more flexible manner depending on the availability of resources in respective countries.

It is now convincing that palm oil products are attractive renewable resources for the production of PHA. In order to ensure sufficient supply of this raw material for large-scale production of PHA, we should first take into account the amount of palm oil products generated in Malaysia every year. The data should include the net balance of global trade (import and export) and existing demand of end users. It was estimated that to produce 5,000 tons of PHA annually, approximately 7,000 tons of soybean oil were required with more than 150 batches of fermentation run (Akiyama et al. 2003). At the moment, this set of simulation is by far the closest and most relevant to the PHA fermentation from palm oil. Over the past decade, the average yield of palm oil on plantations has been measured at 3.8 tons per ha (Murphy 2007). Therefore, in order to produce 1 million tons of PHA, 450,000 ha of oil palm plantation is required. Hence, more lands have to be reclaimed to pave the way for new oil palm plantation in order to produce more oil. Nevertheless, it has been suggested that the yield of oil could be doubled by adopting more efficient management systems coupled with new technologies on existing plantations without the need to expand into new forest lands (Murphy 2004). These include reducing spoilage during transport and storage, hiring of good manager/husbandman, adopting more efficient processing in mills and using high-yield germplasm (Chan et al. 2003). With these strategies, it is possible to increase the yield of palm oil more than fivefold to reach over 20 tons per ha (Murphy 2007).

When the production of PHA by palm oil adheres closely to the proposed fermentation strategy as mentioned by assuming that fermentation conditions and the yield of PHA are comparable, it could be confidently said that approximately 1 million tons of PHA can be produced with the excess supply of 1.7 million tons of palm oil in Malaysia alone (Lam et al. 2009). In the meantime, it is also necessary to take into consideration the surplus amount of palm kernel oil and palm oil by-products in Malaysia that can be potentially converted into PHA, which is a sure sign that the amount of PHA generated could be more. In addition, it had been demonstrated that constant quality of PHA could be achieved even when different palm oil products and by-products were used (Loo et al. 2005). This is of particular importance for industrial-scale production as reliance on a single type and source of the palm oil can be avoided. Although palm oil products are highlighted as promising substrates for commercial production of PHA, to date the development of large-scale production of PHA from palm oil has not been established.

However, with the current knowledge and technology, even if we managed to convert the palm oil products into PHA without having any loss, the amount of PHA materials produced (~1 million tons) is still far below the required volume in 2010. A rough estimation of the demand for biodegradable plastics in 2010 was about 14 million tons (Sudesh and Doi 2005). Meanwhile, the global consumption of petroleum-based plastics was projected to reach over 230 million tons by 2010 (Snell and Peoples 2009). In order to meet even 10 % of this demand, more than 30 million tons of palm oil products are needed. Besides, the emergence of 'biodiesel concept from palm oil' has added complexity to the current situation and will to a certain extent limit the conversion of palm oil into PHA. Nevertheless, after considering all the aspects, it becomes more convincing that the prospect of PHA production in countries with excessive supply of agricultural products or by-products such as Malaysia with its palm oil supply is encouraging.

However, palm oil producers, especially Malaysia still need to convince the public that they consider seriously the matter of sustainability. They have the conscience to protect the environment and put a perfect balance between economic consideration, environmental issues, and future sustenance. In the context of palm oil as source of edible oil and fermentative substrates for PHA biosynthesis or as far as 'food security' issue is concerned, it is the responsibility of the oil palm industries to ensure sufficient supply of edible oil to meet both current and future demands (Berck and Bigman 1993). If this could be achieved, then we can ensure the uninterrupted supply of palm oil for PHA production without affecting palm oil-based food industries.

4.5 Biosynthesis and Characterization of Various Types of PHA From Palm Oil Products

The type of PHA produced from bacterial fermentation is often structurally related to the carbon source used. Nevertheless, some bacteria are capable of synthesizing PHA from structurally unrelated carbon sources. The complex carbon sources are metabolized via multiple biochemical pathways in the bacterial cell and the resulting precursor intermediates are then polymerized by the PHA synthase (PhaC). The type of biochemical pathways in a microorganism and substrate specificity of its PhaC determine the type of PHA synthesized (Steinbüchel and Lütke-Eversloh 2003; Taguchi and Doi 2004). The most conventional and popular carbon sources used for microbial growth and PHA accumulation are sugars such as glucose, fructose, other saccharides, and fatty acids. Plant oils which are known to produce higher yields of PHA per gram of substrate are now gaining much attention as potential feedstock for PHA production. Palm oil products have been investigated for the production of various types of PHA consisting of short-chain-length (scl-) monomers having 3–5 carbon atoms, medium-chain-length (mcl-) monomers having 6–14 carbon atoms or a combination of both types under controlled laboratory conditions. In all cases, the oil is first hydrolyzed by extracellular lipase and the resulting free fatty acids are taken up by cells and used for cell growth and PHA synthesis.

Preliminary studies on PHA production using palm oil products were carried out during the early 1990s. Majid and co-workers reported their initial attempts to use PO-based cooking oil as the sole carbon source for P(3HB) production using *Alcaligenes* sp. AK201 (Majid et al. 1994). P(3HB) at a concentration of 2 g/L was produced from 3 g/L of oil. Later, Majid and co-workers also used *Erwinia* sp. USMI-20 for the production of P(3HB) using CPO, PO and palm kernel oil (PKO) (Majid et al. 1999). P(3HB) content of 46 wt% with cell dry weight (CDW) of 3.6 g/L was achieved after 48 h of cultivation from 4.62 g/L of CPO. Fukui and Doi also showed that up to 80 wt% of P(3HB) could be synthesized by *C. necator* H16 from PO (Fukui and Doi 1998). Recently, various palm oil products were further investigated for the production of P(3HB) by *C. necator* H16 (Lee et al. 2008). The oils tested include fractions obtained from both mesocarp and kernel of palm fruit. The CDW obtained was in the range of 4.6–5.6 g/L with P(3HB) contents between 67 and 78 wt%. In a separate study, a *C. necator* PHB$^-$4 transformant harboring the PHA synthase gene of *Cupriavidus* sp. USMAA2-4 ($phaC_{USMAA2-4}$) was able to synthesize up to 68 wt% of P(3HB) from CPKO after 72 h of cultivation (Kek et al. 2010). Besides *C. necator*, other strains have also been identified to produce P(3HB) using palm oil products. Alias and Tan reported that an isolate obtained from palm oil mill effluent (POME) with 80 % similarity to *Burkholderia cepacia* was able to accumulate 50 and 43.6 wt% of P(3HB) from CPO and PKO, respectively (Alias and Tan 2005). A similar isolate *Burkholderia* sp. USM (JCM15050) was also reported to produce P(3HB) using CPO, PO, CPKO, and palm stearin (PS) (Chee et al. 2010). Up to 70 wt%

of P(3HB) was produced from CPKO. In another study, a fresh water isolate *Chromobacterium* sp. USM2 was able to utilize CPKO for growth and P(3HB) accumulation (Bhubalan et al. 2010b). Cell biomass of 3.0 g/L with 23 wt% of P(3HB) accumulation was observed. To this end, the model PHA strain, *C. necator* H16 seems to be the preferred strain because of its robustness as well as its ability to efficiently convert palm oil into PHA.

Besides P(3HB) homopolymer, biosynthesis of P(3HB-*co*-3HV) copolymer had also been carried out using palm oils in the presence of 3-hydroxyvalerate (3HV) precursors. Initially, Majid and co-workers tested for the production of P(3HB-*co*-3HV) from mixtures of CPO and precursors such as propionic acid, n-propanol, valeric acid, and n-pentanol using *Erwinia* sp. USMI-20 (Majid et al. 1999). Highest 3HV molar fraction of 47 mol% was obtained through a single feeding of valeric acid when the cells were grown on CPO. Recently, Lee and co-workers reported the production of P(3HB-*co*-3HV) copolymers with 3HV molar fraction in the range of 3–10 mol% by adding sodium propionate or sodium valerate with different palm oil products (Lee et al. 2008). High P(3HB-*co*-3HV) accumulation of 90 wt% was obtained from 7.5 g/L of cell biomass when CPKO and sodium propionate were fed to shaken-flask cultures of *C. necator* H16. They investigated the culture parameters such as nitrogen source and its concentration, carbon-to-nitrogen ratio (C/N) as well as pH in order to obtain high CDW and copolymer accumulation. Urea was selected as the most suitable nitrogen source with regard to high copolymer concentration produced. Mixtures of CPKO and 3HV or 4-hydroxybutyrate (4HB) precursors resulted in the production of P(3HB-*co*-3HV) and poly(3-hydroxybutyrate-*co*-4-hydroxybutyrate) [P(3HB-*co*-4HB)] copolymers, respectively (Kek et al. 2010). In their study, *C. necator* PHB⁻4 transformant harboring ($phaC_{US-MAA2-4}$) was able to accumulate up to 31 and 10 mol% of 3HV from sodium valerate and sodium propionate, respectively. The copolymer content varied from 34 to 73 wt% depending on the amount of precursor fed. In the presence of γ-butyrolactone or 4-hydroxybutyrate, P(3HB-*co*-4HB) copolymer with 4HB molar fractions of about 5 mol% was produced.

Biosynthesis of mcl-PHA using palm oil has also been reported. Tan and co-workers investigated the use of saponified palm kernel oil (SPKO) for mcl-PHA production using *Pseudomonas putida* PGA1 (Tan et al. 1997). The mcl-PHA produced mainly contained 3-hydroxyoctanoate (a monomer with eight carbon atoms) as the prominent component. The mcl-PHA contents ranged from 19 to 37 wt% with cell biomass values of 3.0–8.8 g/L. Recently, the same strain and carbon source was investigated in a fermenter-scale experiment (Annuar et al. 2007). Total cell biomass of 2.1 g/L with up to 70 wt% of mcl-PHA was obtained at 12 h of fed-batch cultivation. In a separate study, *P. aeruginosa* was cultivated with PO for the simultaneous production of PHA and rhamnolipids (Marsudi et al. 2008). Here, 36 wt% of mcl-PHA consisting of a mixture of monomers with 6–14 carbon atoms was synthesized. Besides scl-PHA and mcl-PHA, a hybrid polymer of scl-mcl-PHA was also produced from palm oil products using *C. necator* transformants. Fukui and Doi reported

the production of P(3HB-co-3HHx) with 4 mol% of 3HHx using a transformant strain of *C. necator* PHB⁻4 harboring the PHA synthase gene of
A. caviae (*phaC*$_{Ac}$) grown on PO (Fukui and Doi 1998). P(3HB-co-3HHx) content of 81 wt% was achieved from 3.6 g/L of CDW.

In a separate study, a similar transformant was investigated for the production of P(3HB-co-3HHx) copolymer using different palm oil products (Loo et al.
2005). P(3HB-co-3HHx) copolymer with almost constant 3HHx molar fraction
of 5 mol% was synthesized from CPO, PO, and CPKO. The CDW and P(3HB-
co-3HHx) content ranged from 3.1 to 4.3 g/L and 54 to 87 wt%, respectively.
CPKO resulted in higher CDW (4.3 g/L) and copolymer accumulation (87 wt%).
Recently, P(3HB-co-3HHx) copolymers with 3HHx molar fraction in the range of
1–10 mol% were produced using *C. necator* transformant harboring the PHA synthase gene of *Chromobacterium* sp. USM2 (*phaC*$_{Cs}$). Up to 81 wt % of copolymer and 8.1 g/L of CDW were produced from 12 g/L of CPKO (Bhubalan et al.
2010a).

By adding sodium valerate or sodium propionate, a terpolymer consisting of
3HB, 3HV, and 3HHx monomers was produced using CPKO by *C. necator* transformants harboring *phaC*$_{Ac}$ (Bhubalan et al. 2008) and *phaC*$_{Cs}$ (Bhubalan et al.
2008; Bhubalan et al. 2010a, b). The 3HV molar fraction in P(3HB-co-3HV-
co-3HHx) terpolymer was controlled by varying the feeding time and
concentration of 3HV precursor. The transformant expressing *phaC*$_{Ac}$ was found
to synthesize 3HV monomer in the range of 2–60 mol%, whereas the 3HHx molar
fraction was between 2 and 7 mol% (Bhubalan et al. 2008). Sodium valerate was
a better 3HV precursor compared to sodium propionate. The highest 3HV molar
fraction with regard to both high CDW (7.1 g/L) and PHA content (80 wt%) was
35 mol%, which was obtained by feeding 8 g/L of sodium valerate at 48 h of cultivation (Bhubalan et al. 2008). On the other hand, *C. necator* transformant harboring *phaC*$_{Cs}$ produced P(3HB-co-3HV-co-3HHx) with higher and a wider range of
3HV fraction (Bhubalan et al. 2010a). The PhaC of *Chromobacterium* sp. USM2,
which showed higher affinity towards 3HV compared to that of *C. necator*, resulted
in the production of terpolymers with 3HV fractions ranging from 2 to 91 mol%.
The 3HHx molar fractions in the terpolymers (2–9 mol%) were almost similar in
both the transformants (Bhubalan et al. 2010a).

The copolymers and terpolymers produced from the co-feeding of palm oils
and 3HV precursors exhibited improved thermal and mechanical properties compared to P(3HB) homopolymer (Table 4.1). P(3HB-co-3HV) with 3HV molar
fractions ranging from 3 to 10 mol% exhibited melting temperatures (T_m) between
165 and 171 °C (Lee et al. 2008). Meanwhile, P(3HB-co-5 mol% 3HHx) copolymers produced from various palm oil products possessed a much lower T_m in the
range of 125–155 °C (Loo et al. 2005). The results showed that the incorporation of a smaller amount of 3HHx second monomer is more effective in lowering
the T_m compared to the incorporation of double the amount of 3HV. On the other
hand, terpolymers of P(3HB-co-60 mol% 3HV-co-2 mol% 3HHx) and P(3HB-
co-85 mol% 3HV-co-1 mol% 3HHx) recorded even lower T_m values of 81–89 °C
(Bhubalan et al. 2008; Bhubalan et al. 2010a). The glass transition temperatures

Table 4.1 Thermal and mechanical properties and molecular weights of some PHA copolymers and terpolymers produced from palm oil products in the presence of suitable precursor substrates

Monomer composition (mol%)			Thermal properties		Molecular weight		Mechanical properties		Reference
3HB	3HV	3HHx	T_m^a (°C)	T_g^b (°C)	M_n^c ($\times 10^5$)	M_w/M_n^d	Tensile strength (MPa)	Elongation at break (%)	
97	3	0	168	−0.9	6.7	3.9	N.D.	N.D.	(Lee et al. 2008)
93	7	0	166	−0.8	7.4	3.8	N.D.	N.D.	(Lee et al. 2008)
95	0	5	125, 143	−3	4.6	3.3	N.D.	N.D.	(Loo et al. 2005)
91	2	7	144	−3.4	2.2	3.9	22	312	(Bhubalan et al. 2008)
69	24	7	129, 139	−0.8	2.0	2.0	20	710	(Bhubalan et al. 2010a)
38	60	2	81, 87	−13	5.7	4.3	14	421	(Bhubalan et al. 2008)
14	85	1	89	−16.1	N.D.	N.D.	14.5	78	(Bhubalan et al. 2010a)

[a]Melting temperature
[b]Glass transition temperature
[c]Number-average molecular weight
[d]Polydispersity index

(T_g) of terpolymers were generally lower than those of copolymers. Incorporation of both 3HV and 3HHx monomers resulted in the production of flexible yet strong materials (Bhubalan et al. 2010a; Bhubalan et al. 2008). Terpolymer with 24 mol% 3HV and 7 mol% 3HHx exhibited a tensile strength of 20 MPa and elongation at break of 710 %, which is similar to that of common low-density polyethylene (LDPE) (Bhubalan et al. 2010a). Nevertheless, a controlled composition of both 3HV and 3HHx monomers is required in order to obtain a terpolymer with superior properties. The molecular weights of PHA produced from palm oils were generally above 1×10^5 Da, and can be as high as 9.9×10^5 Da. The polydispersity of the polymers ranges from 1.8 to 4.3.

The results presented above show that palm oil products are suitable substrates for the production of various PHAs. In some cases, a precursor substrate such as propionic acid is required to generate 3HV second monomers. Such precursors can be derived from the anaerobic fermentation of POME (Hassan et al. 1996, 1997b), which makes it bio-based and renewable. In general, palm oil products support good cell growth and PHA accumulation. Amongst them, CPKO, which is the oil obtained from the kernel of the oil palm fruit produced the highest PHA yields. Unlike PO, CPKO is mainly used for nonedible purposes by the oleochemical industries because of its higher contents of saturated fatty acids such as dodecanoate (12C) and tetradecanoate (14C).

4.6 Evaluation of Palm Oil By-Products and Spent Cooking Oil as Carbon Source

Global demand for products derived from palm oil has resulted in the generation of large quantities of by-products from its refining processes as well as the wastewater from the oil palm milling facilities. Several studies have been conducted using agricultural and industrial by-products/waste streams as raw materials for various biotechnological processes. Results show that these wastes can be considered as reusable carbonaceous substrates for the production of various value-added materials. Similar efforts were also taken in PHA production. To date, inexpensive waste streams such as whey (Koller et al. 2007; Pantazaki et al. 2009; Povolo et al. 2010), waste oily substrates (Ashby and Solaiman 2008; Kek et al. 2008; Kawata and Aiba 2010), and other industrial wastewater (Pozo et al. 2002; Zakaria et al. 2010a) have been investigated as feedstock for PHA production. These products are assimilated and successfully converted into PHA by using numerous PHA producing strains. Adding to the list above are also palm oil by-products such as PAO and PKAO.

Wild-type *C. necator* H16 was used by Kek and co-workers for the production of P(3HB) using two major by-products rich in free fatty acids from palm oil chemical refining processes; PAO and PKAO, as the sole carbon source (Kek et al. 2008). A total of 5 g/L of these by-products resulted in approximately

4.5 g/L of cell biomass containing 43 wt% P(3HB) after 48 h of cultivation. Results from shake flask cultures have demonstrated that lower urea concentration and elevated culture volume supported better growth and P(3HB) accumulation. A maximum of 5.3 g/L of CDW with 77 wt% of P(3HB) were attained when PKAO was fed as the sole carbon source. Subsequent works by the same group demonstrated the ability of *C. necator* PHB⁻4 transformant harboring *phaC*$_{Cs}$ to biosynthesize P(3HB-*co*-3HHx) copolymer from by-products such as PAO, PKAO, and PFAD. At a concentration of 5 g/L, PAO, PKAO, and PFAD gave comparable CDW (4.0 g/L) and copolymer content (50 wt%) after 72 h of cultivation. In all cases, the 3HHx molar fraction in the copolymer was approximately 5 mol%. Alteration in the C/N ratio of the culture medium showed that 9 g/L of PKAO contributed to the best P(3HB-*co*-3HHx) biosynthesis. Approximately, 69 wt% of copolymer were obtained from 5.5 g/L of cell biomass.

Besides the by-products from palm oil refining processes, the feasibility of using POME as the feedstock in PHA production had also been studied. Anaerobically, treated POME was used as carbon source for PHA production by *Rhodobacter sphaeroides* IFO 12203 and *Comamonas* sp. EB172 (Hassan et al. 1996, 1997b; Zakaria et al. 2010a). The organic acids produced from anaerobic treatment of POME, particularly acetic and propionic acids, were successfully converted into PHA with polymer contents ranging from 59 to 67 wt% of CDW.

Palm oil-based cooking oil comprises mainly of PO and certain amount of other additives. Domestic spent cooking oil is usually discarded after single or multiple uses. As a result of the hydrolysis, thermal oxidation and polymerization reactions during frying at elevated temperature and prolonged periods, used frying oil generally consists of 70 % triacylglycerol (TAG) (fresh cooking oil consists of 95 % TAG) while the remaining fraction consist of oil degradation products (Weiss 1983; Rincon et al. 2010). A range of new polar compounds such as oligomeric TAG or polymers, diacylglycerol, monoacylglycerol, free fatty acids, aldehydes, and ketones are formed, and collectively these are called total polar materials (TPM) (Yates and Caldwell 1993; Takeoka et al. 1997; Dobarganes et al. 2000). It is important that these oils are discharged in a proper manner to avoid pollution and better if they can be reused for other applications. This will help to reduce oil wastage and environmental pollution. Annually, more than 50,000 tons of PO-based spent cooking oils are produced in Malaysia alone (Loh et al. 2006). Domestic spent oils generated by discontinuous fryers contain approximately 10.5–42.1 wt% TPM, whereas those from restaurants and fried food outlets have TPM values ranging from 3.1 to 61.4 wt% (Dobarganes and Márquez-Ruiz 1998). Chemical deterioration of spent cooking oils could be also indicated by the increase in their anisidine value, peroxide value, viscosity, total acid number, and free fatty acid content (Cuesta et al. 1991). The fatty acid composition of waste frying palm oil (WFPO) is shown in Table 3.1. With proper biotechnological processes, these surplus waste materials could be converted into value-added or eco-friendly materials (Akaraonye et al. 2010).

Table 4.2 Biosynthesis of P(3HB) and P(3HB-*co*-3HHx) from various spent palm oils by *C. necator* H16 and its transformant strain harboring the PHA synthase gene of *Chromobacterium* sp. USM2

Spent palm oil[a]		Dry cell weight[b] (g/L)	PHA content[c] (wt%)	Residual biomass (g/L)	Total PHA (g/L)
Produced under laboratory controlled conditions					
Number of usage[d] (times)					
	0[e]	6.2 ± 0.1	61 ± 2	2.4	3.8
	1	5.7 ± 0.4	63 ± 2	2.1	3.6
	2	6.0 ± 0.2	65 ± 2	2.1	3.9
	3	5.4 ± 0.1	64 ± 5	2.0	3.4
	4	5.5 ± 0.1	66 ± 1	2.5	3.0
	5	5.2 ± 0.4	62 ± 5	1.9	3.3
	6	5.4 ± 0.1	56 ± 2	2.4	3.0
	7	5.4 ± 0.1	66 ± 1	1.8	3.6
	8	5.7 ± 0.2	59 ± 3	2.3	3.4
Frying temperature (°C)	Frying duration (min)				
150	3	6.3 ± 0.1	73 ± 2	1.7	4.6
150	5	5.6 ± 0.2	63 ± 3	2.1	3.5
180	3	3.8 ± 0.1	49 ± 2	1.9	1.9
180	5	5.8 ± 0.3	59 ± 4	2.4	3.4
Spent palm oil collected from cafeteria					
Time profiles[f]	Cultivation time (h)				
	12	1.2 ± 0.1	17 ± 1	1.0	0.2
	24	5.1 ± 0.2	44 ± 1	2.9	2.2
	36	6.8 ± 0.4	62 ± 2	2.6	4.2
	48	9.6 ± 0.5	65 ± 2	3.4	6.2
	60	10.6 ± 0.2	72 ± 3	3.0	7.6
	72	9.9 ± 0.1	67 ± 4	3.3	6.6
	84	9.1 ± 0.4	65 ± 4	3.2	5.9
	96	8.9 ± 0.2	69 ± 4	2.8	6.1

[a]Cells were cultivated on various spent palm oils as the sole carbon source at 30 °C and mineral medium with the initial pH of 7.0
[b]Dry cell weight after lyophilization
[c]PHA content and composition in lyophilized cells were determined by gas chromatography analysis
[d]Palm oil was used to fry nuggets at 120 °C for 5 min and cooled to room temperature before the next round of frying was carried out
[e]Palm oil with zero usage refers to sterile fresh oil
[f]The transformant strain was grown on 10 g/L of spent palm oil and 15 mM of urea at 30 °C with an agitation speed of 200 rpm
DCW and PHA content were determined in triplicate; mean values and standard deviations are provided

Table 4.3 Biosynthesis of PHA from palm oil-based carbon sources using various wild-type and recombinant microorganisms

Carbon substrate(s)	Bacterial strain(s)	PHA content (wt%)	Types of PHA	Co-monomer composition (mol%)			Reference(s)
				4HB	3HV	3HHx	
PO	Alcaligenes sp. AK201	42	P(3HB)	–	–	–	(Majid et al. 1994)
CPKO	C. necator H16 and its PHB^{-4} (phaC$_{USMAA2}$$^{-4}$)	68	P(3HB)	–	–	–	(Kek et al. 2010)
CPO, PO, PKO, PS	Burkholderia sp. FLP1	43–57	P(3HB)	–	–	–	(Alias and Tan 2005)
CPO, PO, CPKO, PS, PAO, PKAO, PFAD	Burkholderia sp. USM	70	P(3HB)	–	–	–	(Chee et al. 2010)
CPKO	Chromobacterium sp. USM2	23	P(3HB)	–	–	–	(Bhubalan et al. 2010b)
PAO, PKAO	C. necator H16 ATCC 17699	58–79	P(3HB)	–	–	–	(Kek et al. 2008)
Spent cooking oil	C. necator H16	49–73	P(3HB)	–	–	–	Unpublished data
PO + 3HV precursors	Erwinia sp. USMI-20	46	P(3HB) P(3HB-co-3HV)	–	6–47	–	(Majid et al. 1999)
CPO, PO, CPKO + 3HV precursors	C. necator H16	64–90	P(3HB) P(3HB-co-3HV)	–	2–23	–	(Lee et al. 2008)
CPKO + 3HV/ 4HB precursors	C. necator PHB^{-4} (phaC$_{USMAA2-4}$)	34–73	P(3HB-co-3HV) P(3HB-co-4HB)	4–5	4–31	–	(Kek et al. 2010)
SPKO	P. putida PGA1	19–37	mcl-PHA	–	–	–	(Tan et al. 1997; Annuar et al. 2007)
PO	P. aeruginosa IFO3924	36	mcl-PHA	–	–	–	(Marsudi et al. 2008)
PO	C. necator PHB^{-4} (phaC$_{Ac}$)	81	P(3HB-co-3HHx)	–	–	4–5	(Fukui and Doi 1998)
CPKO	C. necator PHB^{-4} (phaC$_{Ac}$)	87	P(3HB-co-3HHx)	–	–	5	(Loo et al. 2005)
PAO, PKAO, PFAD	C. necator H16 and its PHB^{-4} (phaC$_{Cs}$)	72	P(3HB-co-3HHx)	–	–	3–6	Unpublished data
Spent cooking oil	C. necator H16 and its PHB^{-4} (phaC$_{Cs}$)	28–72	P(3HB-co-3HHx)	–	–	3–5	Unpublished data
CPKO + 3HV precursors	C. necator PHB^{-4} (phaC$_{Ac}$)	6–79	P(3HB-co-3HV-co-3HHx)	–	2–60	2–7	(Bhubalan et al. 2008)
CPKO + 3HV precursors	C. necator PHB^{-4} (phaC$_{Cs}$)	51–86	P(3HB-co-3HV-co-3HHx)	–	2–91	1–10	(Bhubalan et al. 2010a)

PO palm olein, CPKO crude palm kernel oil, CPO crude palm oil, PKO palm kernel oil, PS palm stearin, PAO palm acid oil, PKAO palm kernel acid oil, PFAD palm fatty acid distillate, SPKO saponified palm kernel oil

Various types of uncharacterized spent cooking oils had been previously investigated for the production of PHA (Taguchi and Doi 2004; Fernandez et al. 2005; Haba et al. 2007; Song et al. 2008). They were found to be a feasible yet inexpensive source of carbon feedstock for PHA synthesis. Recently, Rao and co-workers used mixtures of PO-based spent cooking oil and 1,4-butanediol for the biosynthesis of P(3HB-*co*-4HB) copolymer by *C. necator* (Rao et al. 2010). They reported that P(3HB-*co*-4HB) copolymer with 15 mol% of 4HB was produced and the copolymer showed good biocompatibility to be developed as an absorbable biomaterial.

Besides, a preliminary study employing wild-type *C. necator* was carried out for the production of P(3HB) using similar PO-based spent cooking oil which was generated under laboratory controlled conditions (Table 4.2). The performance of this microorganism was evaluated using spent oil after repeated usage, different frying temperature and duration. It was demonstrated that the increase in usage (one to eight times of frying) of the cooking oil did not have a significant effect on cell biomass and P(3HB) synthesis. The CDW and P(3HB) content were found to be in the range of 5.2–6.0 g/L and 59–66 wt%, respectively, at 48 h of cultivation. When the frying temperature and frying duration was altered, the latter produced greater effect on the growth of *C. necator*. Higher residual biomass (nonPHA cellular material) was observed when using spent oil obtained from frying temperature of 180 °C as compared to 150 °C. The frying duration for both cases was 3 and 5 min, respectively. In a separate study, spent palm oil collected from a cafeteria was used as the sole carbon source for P(3HB-*co*-3HHx) copolymer synthesized by employing *C. necator* PHB⁻4 harboring *phaC*$_{Cs}$. By adding 10 g/L of spent oil and 15 mM of urea, the best CDW of 10.6 g/L and copolymer accumulation of 72 wt% were achieved. The results from the preliminary studies above indicated that PO-based spent cooking oil holds much potential to be used as carbon source for PHA synthesis in large-scale fermentations, which is currently ongoing.

Utilization of readily available and renewable waste/by-products for microbial fermentation can be regarded as promising platform for the development of sustainable PHA production systems. Their usage in the biosynthesis of value-added green material such as PHA can provide palm oil and food industries with a strategy to control pollution and waste disposal management problems. On the other hand, conversion of spent cooking oil into PHA provides a brilliant solution to overcome excessive oil wastage and sustainable use of available resources in the future. A summary of PHA produced from various palm oil products including by-products is shown in Table 4.3. It can be seen that palm oil is indeed a suitable carbon substrate able to be utilized by many types of bacteria for growth and production of various types of PHA. With proper biotechnological processes and resource management, potential application of palm oil products for PHA production can be harnessed to the fullest.

Chapter 5
Jatropha Oil as a Potential Carbon Source for PHA Production

Abstract Studies have shown that the production of polyhydroxyalkanoate (PHA) from plant oils is more efficient than from sugars in terms of productivity. Among the various plant oils, palm oil is the most efficiently produced oil in the world. The main application of palm oil is as a source of dietary fat. The conversion of food grade substrates to non-food materials is of concern because of the increasing need to feed the rapidly growing human population. Therefore, the by-products of the plant oil industry may be a better feedstock for PHA production. Alternatively, non-food grade oils such as jatropha oil can be developed as a feedstock for PHA production. This chapter looks at the potential of jatropha oil as a possible feedstock for the biosynthesis of PHA.

Keywords Crude palm kernel oil (CPKO) • *Jatropha curcas* • Jatropha oil • Palm oil • Plant oils • Polyhydroxyalkanoate (PHA)

5.1 *Jatropha curcas* L. (Linnaeus)

Jatropha curcas L. is a multipurpose plant with many contributions and economic potential in medical and industrial uses. "*Jatropha*" is derived from the *iatros* and *troph*e in Greek with the meaning of doctor and food, respectively (Heller 1996). It is commonly recognized as the physic nut. It is the most primitive form of *Jatropha* genus, which belongs to Euphorbiaceae family. Most of the *Jatropha* species are originally from America except 66 species which are of Asian (*Jatropha villosa, Jatropha multifida, Jatropha podagrica, Jatropha gossypiifolia,* etc.) and African origin (*Jatropha afrocurcas, Jatropha macrophylla,* etc.) (Heller 1996).

Jatropha curcas L. is a small perennial shrub originally from Caribbean, most probably is from Mexico. It is believed to have been transported by Portuguese sailor to Africa and Asia and cultivated for the oil in hedges (Heller 1996). After decades of distribution, *J. curcas* L. can be found in the tropical and subtropical regions around the world. This easy-growing and multipurpose plant exists in regions with various vernacular names (Table 5.1).

K. Sudesh, *Polyhydroxyalkanoates from Palm Oil: Biodegradable Plastics,* 63
SpringerBriefs in Microbiology, DOI: 10.1007/978-3-642-33539-6_5,
© The Author(s) 2013

Table 5.1   Various vernacular names of *J. curcas* L. (Heller 1996)

Regions	Vernacular names
French	Pourghère, pignon d'Inde
Dutch	Purgeernoot
Portuguese	Purgueira
Mexico	Piñoncillo
India	Bagbherenda, jangliarandi, safed arand
Nepal	Kadam
Malaysia	Jarak belanda, jarak pagar, jarak keling
Indonesia	Jarak pagar, jarak kosta, balacai
Philippines	Tagumbau-na-purau, tuba, tubing-bakod
Thailand	Ma yao, sabuu dam, salot paa
Cambodia	Kuang, lohong
Laos	Nhao

5.2   The advantages of *J. curcas* L

Jatropha curcas L. became economically important for several reasons. *J. curcas* L. is a well-adaptable crop which is able to grow on marginal areas with unfertile soil and low rainfall (Openshaw 2000). It can grow on sand, stony soil, wasteland, and even in the crevices of rocks except water clogging soil. Its low water requirement allows it to grow in a dry area with only 250 mm of rainfall per annum (Foidl et al. 1996). In order to counter the drought and low nutrient content of the soil, it sheds its leaves to reduce the transpiration loss and to improve the fertility of the soil (Kumar and Sharma 2008). *J. curcas* L. can tolerate high temperature but not frost. Therefore, it can be found mostly in tropical and subtropical areas. The tap roots and the lateral roots of this plant are well developed. The tap root of this plant is thin and can grow into deeper layers of soil to obtain nutrients without harsh competition with other crops (Achten et al. 2008). Due to the above advantages, *J. curcas* L. is planted to reclaim wasteland and to prevent soil erosion (Jain and Sharma 2010).

J. curcas L. is a large shrub with the height of 5–7 m. It is a non-edible plant because the stems, leaves, and seeds are toxic to animals. As animals will not browse it, it was planted as a hedge or live fence along the roadsides, railway tracks, and around some crop plantations (Heller 1996). Live fence is more cost-effective than wire fence. In addition, *J. curcas* L. is less threatened by diseases, insects, and fungal pests than other agricultural crops. It can be planted to protect some crops except if it has common pests with the crops. It also helps in providing humus and improving microclimate of the crop fields (Openshaw 2000).

J. curcas L. is a perennial plant with a productive life of 30–50 years. It starts to bear fruits after two years of cultivation, and the best production starts from the fourth year (Kumar and Sharma 2008). Each fruit contains 2–3 oblong black seeds (Salimon and Abdullah 2008). The seeds contain 40–60 % of oil which is used for many purposes except for consumption (Martínez-Herrera et al. 2006; Salimon and Abdullah 2008). Jatropha oil is used as a traditional medicine for inducing hair growth, treatment of skin diseases, and soothing pain (Gübitz et al. 1999; Makkar and Becker 2009). Jatropha oil is nonedible and may be used as pesticide or insecticide. For industrial usages, it is used in soap production and cosmetic manufacture. Recently, jatropha oil is gaining a lot of attention because of its potential for biodiesel production (Berchmans and Hirata 2008; Ganapathy et al. 2009; Lim and Teong 2010). This renewable resource is proposed to reduce the greenhouse gas emission and the dependence on fossil fuel. The seed cake, a by-product of oil extraction, could be used as fertilizer because it is rich in nitrogen, phosphorus, and potassium. It also has potential as feedstock for biogas production and methane production (Kumar and Sharma 2008). The seed cake can be a good source of protein for animal feed after detoxification because it contains up to 64 % of protein and high level of most essential amino acids, which are even higher than that of the FAO reference protein (Aregheore et al. 2003; Makkar and Becker 1997). In Mexico, some varieties of *J. curcas* are found to be nontoxic and can be used as animal feed without detoxification (Makkar et al. 1998; Martínez-Herrera et al. 2006).

All parts of *J. curcas* L. is used in medical and veterinary purposes. The seeds, leaves, and bark contain toxins and phytochemicals, which have molluscidal and insecticidal properties (Rug et al. 1997; Rahuman et al. 2008). Various phytochemicals of *J. curcas* L. were identified and studied for their antitumor (Balaji et al. 2009), anti-inflammatory (Mujumdar and Misar 2004), antimicrobial (Ravindranath et al. 2004), antiviral (Matsuse et al. 1998), and wound healing activity (Nath and Dutta 1992; Esimone et al. 2008). The bioactive compounds are not only used in medical and veterinary but could also be used in agriculture.

5.3   Toxins and Phytochemicals of *J. curcas* L

Seed, seed oil, and seed cake of *J. curcas* L. are toxic to human and animals. Although the seed meal is rich in protein, it is reported to be toxic to rats (Goonasekera et al. 1995; Makkar and Becker 1999), mice (Li et al. 2010), goat (Adam and Magzoub 1975), fish (Makkar and Becker 1999), and human. The patients who accidentally consumed jatropha seeds showed symptoms of vomiting, diarrhea, and giddiness (Martínez-Herrera et al. 2006). A general antitoxin is suggested as the effective treatment (Ye et al. 2009). Phorbol esters that are present at a high concentration in jatropha seeds are responsible for the toxicity effect. A phorbol-type diterpene, Jatropherol-I (JaI), was reported to have damaging effects on the tissue structure in the digestion system of silkworm (Jing et al. 2005). Generally, phorbol esters are carcinogenic, tumor promoters, and highly irritant (Goel et al. 2007). The phorbol esters contain tetracyclic diterpene basic skeleton known as tigliane (four rings) and the fatty acids which form ester bonding with the hydroxyl groups in tigliane. The phorbol esters mimic the diacylglycerol to activate the protein kinase C, which plays important roles in signal transduction for various biochemical reactions in cells. Hyper activation of protein kinase C may assist in expression of oncogene induced by a carcinogen and causing the promotion of tumor (Goel et al. 2007; Mosior and Newton 1995; Poullin et al. 1987). The phorbol esters act as cocarcinogens. In some reports, phorbol esters of *J. curcas* L. also show insecticidal and molluscidal activities (Rug et al. 1997; Wink et al. 1997). A type of *Jatropha* phorbol ester, 12-deoxy-phorbol-13-phenylacetate, was reported to have anti-HIV activity and inhibited HIV from entering targeted cells (Wender et al. 2008).

The seed contains curcin (a lectin of *J. curcas* L.), a ribosome inactivating protein. It is toxic to most vertebrates as it inhibits the synthesis of protein, but it is potential for immunotoxins preparation to treat cancer and autoimmune diseases (Lin et al. 2003). Curcin 2 extracted from stressed leaves exhibits antiviral and antifungal activity and reduces the damages of plants (Huang et al. 2008). A high level of phytate in jatropha meal might react with minerals and protein to form complexes which decrease the bioavailability and digestibility (Reddy and Pierson 1994; Harland and Morris 1995; Lan et al. 2010). Trypsin inhibitor activities were also detected in jatropha meal. Trypsin inhibitor interferes with pancreatic proteolytic enzymes and thus reduces the protein digestibility leading to severe growth depression (Kumar et al. 2003; Clarke and Wiseman 2005).

5.4 Jatropha Oil

Jatropha oil which refers to *J. curcas* oil is extracted from *J. curcas* L. seed by using mechanical or chemical extractions (Aderibigbe et al. 1997; Forson et al. 2004). For mechanical extraction method, the whole seed or kernel alone or the mixture of both can be subjected to manual or engine presses. In order to obtain higher yield for commercial applications, chemical extraction using solvent is recommended, but it is applied only to kernels. Solvent extraction is time-consuming and has greater environmental impacts. The aqueous enzymatic extraction assisted by sonication could help in reducing these problems (Shah et al. 2005).

Jatropha oil contains saturated fatty acids and unsaturated fatty acids, mainly palmitic acid, stearic acid, oleic acid, and linoleic acid (Akintayo 2004; Martínez-Herrera et al. 2006). Oleic acid is the predominant fatty acid in jatropha oil, about 42 % in fatty acid composition (Table 5.2). Therefore, jatropha oil is defined as oleic oil. Due to the high percentage of unsaturated fatty acids, jatropha oil possesses semi-drying property, which makes it a potential substance for surface coating and low pour point biodiesel (Augustus et al. 2002; Salimon and Abdullah 2008). The high viscosity of jatropha oil is due to its large molecular mass but the chemical structure of oil does not affect the combustion performance in a compression ignition engine after the blending with diesel (Pramanik 2003). In fact, the viscosity of jatropha oil adds to the advantage in production of soap, candles, and cosmetic products (Kumar and Sharma 2008; Openshaw 2000). The jatropha oil has high saponification and iodine value (Akintayo 2004). It is suitable to be used for making soaps, shampoos, varnishes, shoe polish, and alkyd resin.

Table 5.2 Fatty acids composition of jatropha oil

Fatty acid	Composition (%)
Saturated	
Lauric $C_{12:0}$	Trace*
Myristic $C_{14:0}$	0.10 ± 0.00
Palmitic $C_{16:0}$	17.10 ± 0.05
Stearic $C_{18:0}$	4.30 ± 0.02
Arachidic $C_{20:0}$	Trace*
Unsaturated	
Palmitoleic $C_{16:1}$	1.20 ± 0.00
Oleic $C_{18:1}$	42.00 ± 0.02
Linoleic $C_{18:2}$	34.80 ± 0.10
Linolenic $C_{18:3}$	0.10 ± 0.00

*Trace: concentration less than 0.01 %

Jatropha oil also contains compounds that cannot be saponified, which are organic soluble compounds comprising mainly the polyphenols. In some studies, a variety of sterols and terpene alcohols were found in jatropha oil. In different samples of jatropha oil which were extracted from the *J. curcas* planted in Nigeria, different sterols including campesterol, stigmasterol and β-sitosterol, and terpene alcohols including dihydroparkeol and taraxasterol were detected (Adebowale and Adedire 2006; Akintayo 2004). The type of compounds found might vary in different samples depending on the cultivation environments and genetics of the plants (Kaushik et al. 2007). Besides sterols and terpene alcohols, phorbol esters are also found in jatropha oil. During the oil extraction, as high as 75 % of the phorbol esters in *Jatropha* fruits might be transferred into the jatropha oil. Hence, phorbol esters are suggested to be the main compound responsible for toxicity effect in jatropha oil (Makkar et al. 2008). However, some genotypes of *J. curcas* in Mexico may contain very low phorbol esters content and thus making the seeds and oil nontoxic (Martínez-Herrera et al. 2010).

5.5 Case Study on the Use of Jatropha Oil as a Carbon Substrate for PHA Biosynthesis

Research on *J. curcas* L. as a renewable resource has been carried out intensively because this plant is claimed to thrive on even the poorest stony soil and in very harsh climates. More importantly, the cultivation of *Jatropha* plants is estimated to cost less than soybeans (0.39 and 1.64 USD/kg oil, respectively) due to their lower fertilizer and pesticide requirements (Gui et al. 2008). In addition, *J. curcas* has a productive life of up to 50 years, and the seeds may yield up to 58 wt% oil (Martínez-Herrera et al. 2006). Apart from the advantages mentioned above, it is speculated that jatropha oil has great potential as the feedstock for PHA production by bacterial fermentation in Malaysia based on several reasons. First, Malaysia has the potential to cultivate *Jatropha* plantations because this plant grows more rapidly and produces more seeds in the tropics, especially on land with sufficient nitrogen and rainfall (Openshaw 2000). Second, it was reported recently that the *Jatropha* seeds from the northern part of Malaysia have high lipid content of about 60 wt% of oil (Salimon and Abdullah 2008).

Since the announcement of the 8th Malaysia Plan for year 2001–2005, renewable energy has been promoted as the sustainable fifth fuel other than the four fuel-mix energy supply consisting of oil, natural gas, coal, and hydropower (EIBM 2006). The main reason for promoting the renewable energy is to reduce the dependence on conventional energy resources. In Malaysia, palm oil is the main renewable resource especially for biodiesel production. In addition to palm oil, *Jatropha* oil has been taken into consideration as supplementary renewable energy in order to balance the renewable energy supply and exportation (Lim and Teong 2010). Due to the hardy and easy-growing behavior of *J. curcas*, its plantation on marginal land is encouraged by Malaysia Rubber Board, which is now responsible for *Jatropha* research. By cultivating on marginal land, which is not suitable for crop plantation, the efficiency of land utilization can be increased without raising the conflict of using fertile land for fuel. Besides, using the non-edible jatropha oil, which is not a food grade oil as an energy resource does not interfere with the global food supply. Realizing the increasing potential of *Jatropha* as renewable energy, many start-up companies are now investing in its research and cultivation. By the end of 2010, the land cultivation area was expected to reach 1 million ha (Lim and Teong 2010). Many studies have been done on the production of biodiesel from jatropha oil (Ganapathy et al. 2009; Lu et al. 2009; Pramanik 2003) and the price is expected to be lower than that of rapeseed, soybean, palm oil, and waste cooking oil (Lim and Teong 2010). However, the utilization of jatropha oil as a carbon feedstock for the production of bioplastics by bacteria was yet to be proven until quite recently. The potential of jatropha oil as a carbon source for PHA production using *C. necator* H16 was investigated and reported by Ng et al. (2010).

C. necator is one of the most productive hosts for PHA biosynthesis. Its physiological and biochemical mechanisms are also well studied. It is the bacterium that was chosen for the commercial production of P(3HB-*co*-3HV) copolymer by Imperial Chemical Industries (ICI) under the trademark Biopol® (Luzier 1992; Braunegg et al. 1998). The accumulation of PHA in the cells is stimulated by the limitation of elements such as nitrogen and phosphorus under the conditions of excess carbon source. Among all the carbon sources that have been tested for PHA production, many studies have shown that the renewable plant oils, with higher carbon content than sugars, are efficient carbon sources for producing high yield of PHA (Akiyama et al. 2003). By using plant oils such as olive oil, corn oil, and palm oil as the carbon sources, the *C. necator* strain can accumulate P(3HB) homopolymer up to 80 wt% of the CDW (Fukui and Doi 1998). Besides P(3HB), a copolymer containing 3HV units with high molecular weight can also be synthesized from the mixture of plant oils and 3HV precursors by *C. necator* H16 (Lee et al. 2008). Based on the results obtained for palm oil, *C. necator* H16 was chosen to be the host for evaluating the efficacy of jatropha oil to support the cell growth and PHA production.

Nitrogen-limiting conditions are more effective than other nutrient depletion in stimulating the biosynthesis of P(3HB) in *C. necator* H16 (Shi et al. 1997). As the P(3HB) biosynthesis in *C. necator* H16 is partially growth associated (Shahhosseini 2004), the selection of a suitable nitrogen source which is utilized for cell growth seems to be essential. Kek et al. (2008) reported that during the 48 h of cultivation, *C. necator* H16 preferably utilized ammonium chloride and urea that resulted in the highest P(3HB) content and CDW, respectively, in the presence of palm acid oil (PAO) while sodium nitrate resulted in the highest CDW and P(3HB) content in the presence of palm kernel acid oil (PKAO). The effects of different nitrogen sources on the growth and P(3HB) biosynthesis seem to be varied in the presence of oils with different fatty acids composition. The effects of different nitrogen sources on P(3HB) biosynthesis using jatropha oil as the sole carbon source was in accordance with a previous research (Lee et al. 2008). It was found that different nitrogen sources affected both cell biomass and PHA biosynthesis. This might be due to the effect of fatty acids and nitrogen sources on the transportation mechanism and permeability of the membrane (Ng et al. 2010). The acidic substances, for example, the short-chain fatty acids, can disrupt the proton gradient across the cell membranes thus activating the transportation of excessive protons in order to maintain the pH in the cytoplasm (Defoirdt et al. 2009). The maintenance of the pH consumes cellular energy which is needed for microbial activities including cell growth and PHA biosynthesis. The proton gradient may affect the uptake of nitrogen compound having different charges. It was shown that the acidic $NH_4H_2PO_4$ (pK_a 3.8–4.4) carrying two hydrogen atoms resulted in the lowest CDW and total P(3HB). Sodium nitrate and urea which are almost neutral (pK_a 7.0–7.2) resulted in better biomass and P(3HB) production compared to the other nitrogen sources. However, urea was chosen as the most suitable nitrogen source in the subsequent experiments because of its lower price and the high productivity of P(3HB) (Kek et al. 2008). As with other complex nitrogen sources, e.g., peptone and yeast extract, urea significantly increased both cell biomass and PHA production (Khanna and Srivastava 2005b; Sabra and Abou-Zeid 2008).

The effects of urea concentration (ranging from 0.18 to 1.2 g/L) on P(3HB) biosynthesis by *C. necator* H16 from jatropha oil were evaluated. The cultivation without supplementation of urea was used as the control because some cell growth may occur because of the residual nitrogen source in the medium, which was transferred together with the preculture. *C. necator* H16 accumulates P(3HB) as carbon and energy storage under limited nitrogen conditions (Sudesh et al. 2000). However, a very low concentration of urea limits the cell growth, thus lowering total P(3HB) productivity even though high content of P(3HB) is accumulated in the viable cells. During the limitation of nitrogen, the amino acid synthesis pathway, especially the transamination reaction of α-ketoglutarate, a precursor in TCA cycle, would be blocked (Shi et al. 1997). Subsequently, the concentration of NADH and NADPH would increase and inhibit the citrate synthase and isocitrate dehydrogenase which are key enzymes in TCA, resulting in the channeling of acetyl-CoA into the P(3HB) biosynthesis pathway instead of the TCA cycle (Anderson and Dawes 1990; Steinbüchel and Lütke-Eversloh 2003; Tsuge 2002; Verlinden et al. 2007). When the urea concentration was increased from 0.18 to 0.54 g/L, the cell growth increased significantly with a slight decrease of P(3HB) content. However, it resulted in a significant increase in total P(3HB). An increase in nitrogen concentration would increase the enzyme activities in TCA cycle to generate energy and amino acids for the cells. The high activity of citrate synthase, which condenses the acetyl-CoA and oxaloacetate into citrate, increases the free CoASH which then inhibits the condensation of 3-ketothiolase to acetoacetyl-CoA for P(3HB) biosynthesis. However, the inhibition reaction might be limited leading to a slight decrease of P(3HB) accumulation (Du et al. 2001a). A more significant decrease in P(3HB) content could be observed when the concentration of urea was more than 0.66 g/L. The cell growth was saturated when the concentration of urea was more than 0.54 g/L with decreasing P(3HB) accumulation. Therefore, urea at 0.54 g/L was found to be the most suitable concentration for high CDW and P(3HB) accumulation.

In the study on the effects of jatropha oil concentration on P(3HB) biosynthesis, an increase in the concentration of jatropha oil also increased the C/N ratio (Ng et al. 2010). Both the ratio and concentration of carbon and nitrogen sources are important factors that affect P(3HB) biosynthesis and cell growth (Lee et al. 2008; Patnaik 2006). A high PHA accumulation could be achieved by maintaining the C/N ratio at its optimal, which falls in the range of 20–50 (Amirul et al. 2008; Lee et al. 2008). As the concentration of jatropha oil increased, the increase in P(3HB) accumulation was more significant than the cell growth. When plant oil is supplemented into the medium, C. necator H16 secretes extracellular lipase to catalyze oil hydrolysis and release free fatty acids (Fukui and Doi 1998; Kahar et al. 2004). The free fatty acids are then transferred into the cells and converted into respective acyl-CoA thioesters by thiokinase before they are oxidized repeatedly in the fatty acid β-oxidation cycle to form many units of acetyl-CoA (Steinbüchel and Lütke-Eversloh 2003; Sudesh et al. 2000). The acetyl-CoA may enter the TCA cycle for the metabolism or P(3HB) biosynthesis pathway. However, the limitation of nitrogen would block the TCA cycle and restrict the cell growth. The best C/N ratio determined in previous studies was about 38 at 12.5 g/L of jatropha oil, which was efficiently utilized by the cells, leaving almost negligible residual oil.

Unlike most edible vegetable oils, the non-edible jatropha oil contains several toxic compounds that are harmful to human and most mammals (Gandhi et al. 1995; Jing et al. 2005). The stems, leaves, and roots of Jatropha sp. such as Jatropha maheshwarii (Viswanathan et al. 2004) and Jatropha gaumeri (Can-Aké et al. 2004) were reported to possess antimicrobial activity. The extraction of various parts of J. curcas was found to contain several types of phytochemical such as steroids, tannins, alkaloids, flavonoids, phorbol esters, etc. (Martínez-Herrera et al. 2006; Igbinosa et al. 2009). The biological activity of these phytochemicals act together to exhibit profound antimicrobial effect. Despite these, there are few reports of antimicrobial activity study on the oil of J. curcas. Jatropha oil is extracted from the seeds of J. curcas which contain toxic and antinutritional compounds such as the lectins, saponins, phytate, trypsin inhibitor, and the most potent, phorbol esters (Makkar et al. 1998).

Table 5.3 Biosynthesis of P(3HB) from various plant oils by *C. necator* H16

Plant oil[a] (5 g/L)	Total P(3HB) (g/L)	Cell dry weight (g/L)	P(3HB) content[b] (wt%)	References
Jatropha oil	4.4	6.1	72	(Ng et al. 2010)
Crude palm kernel oil	3.4	5.0	67	(Lee et al. 2008)
Crude palm oil	3.5	4.6	75	(Lee et al. 2008)
Palm kernel oil	4.1	5.5	75	(Lee et al. 2008)
Palm olein	3.6	5.2	70	(Lee et al. 2008)
Olive oil	3.9	4.9	80	(Lee et al. 2008)
Sunflower oil	3.4	4.7	72	(Lee et al. 2008)
Coconut oil	3.3	4.4	76	(Lee et al. 2008)
Soybean oil	3.0	3.6	82	(Mifune et al. 2008)

P(3HB) poly(3-hydroxybutyrate)
[a]Incubated for 72 h at 30 °C, initial pH 7.0, 200 rpm
[b]PHA content in freeze-dried cells

Phorbol esters are hydrophobic and soluble in oil. It might be the main constituent in jatropha oil causing the toxic effect (Gandhi et al. 1995; Aregheore et al. 2003; Devappa et al. 2010). The toxins in jatropha oil were found not to affect the P(3HB) biosynthesis by the bacterial strain used in this study even when the concentration of oil was as high as 12.5 g/L. The toxin concentration, most probably the phorbol esters content, was too low to exhibit toxicity effects on the cells (Devappa et al. 2010). This is further supported by the analysis on phorbol esters content in the crude oil of *J. curcas* seeds from Indonesia, India, and Malaysia by (Ahmed and Salimon 2009). The lowest concentration and least types of phorbol esters were detected in jatropha oil from Malaysia.

The production of P(3HB) from jatropha oil is as good as other plant oils with 4.4 g/L of total P(3HB) (Table 5.3). *C. necator* cells utilize palmitic acids (C_{16}), oleic acids ($C_{18:1}$), and linoleic acids ($C_{18:2}$) better than linolenic acids ($C_{18:3}$) for cell growth (Kahar et al. 2004). The jatropha oil (Table 5.2) contains only 0.1 % of linolenic acid with most of the fatty acids present accounted for 93.9 % palmitic acid, oleic acid, and linoleic acid. This oil composition also helps to explain the high yield of biomass and P(3HB) from jatropha oil. In another study, oleic acid, which is the major type of fatty acid contained in jatropha oil, was shown to increase the 3HB yield (Marangoni et al. 2000). Besides, long-chain fatty acids (C_{11} to C_{19}) with higher carbon content than sugars are certainly better substrates for producing high CDW (Akiyama et al. 1992, 2003). Therefore, jatropha oil has been identified as a feasible and excellent carbon source for the production of P(3HB) by *C. necator* H16.

C. necator H16 could utilize plant oil as carbon source after the hydrolysis of oil into fatty acids and glycerol by the secreted extracellular lipases and esterases. Initially, one pair of lipase/chaperone gene and one esterase gene were detected in *Ralstonia* sp. M1 by Quyen et al. (2005, 2007). However, recent gene expression study by Brigham et al. (2010) showed that *C. necator* H16 putatively possessed two lipase genes, one chaperon and one esterase. They proposed that the esterases and some of the lipases of *C. necator* H16 break down the oil droplets into smaller particles in medium, while the lipases hydrolyze the oil to release free fatty acids which then emulsify the oil droplets. The fatty acids are then translocated into the cells to be oxidized. In the presence of some compounds, e.g., fatty acids, triglycerides, and surfactants, lipase secretion can be induced (Kulkarni and Gadre 1999; Boekema et al. 2007). Referring to another study, *Jatropha* seed cake has been proven to be a potential substrate for lipase production (Mahanta et al. 2008). Therefore, to induce lipase secretion for better oil utilization for cell growth and higher P(3HB) production, a small amount of jatropha oil was added into the pre-culture in the study by Ng et al. (2010). Lipase production by *C. necator* H16 increased faster during the initial stage (0–18 h) of cultivation when 1 g/L of jatropha oil was added to the preculture medium. It was suggested that the jatropha oil in the preculture medium activated the cells' lipase operon thereby increasing the lipase production during the early stages of PHA production (Rosenau and Jaeger 2000; Kahar et al. 2004). Despite the enhancement of lipase activity in the initial stage of production, cell biomass and P(3HB) content during the cultivation were not enhanced. The production of P(3HB) did not correlate positively with the enhancement of lipase activity. However, jatropha oil was proven to be a good substance for stimulating the secretion of lipase by *C. necator* H16.

The results in Table 5.3 show that the production of P(3HB) by *C. necator* H16 from jatropha oil was compatible with those from other plant oils. The toxins in jatropha oil did not adversely affect the P(3HB) biosynthesis and accumulation in the cells. In addition, microscopic observations of cells under the phase-contrast microscope, fluorescent microscope, and transmission electron microscope, did not show any morphological defects in the cells at different periods of cultivations with 12.5 g/L of jatropha oil as the sole carbon source. *C. necator* H16 divided actively during the early stage in the presence of 12.5 g/L of jatropha oil. The width of cells during the early stage was about the same as that reported by Tian et al. (0.65–0.75 μm) (2005). The granules were seen to accumulate along the longitudinal center of the rod-shaped body, which was similar to some earlier reports. The localization of the granules seemed to have initiated near the nucleoid region or the dark-stained mediation elements, which appeared at the longitudinal center of the cells (Tian et al. 2005; Kek et al. 2008). Some studies proposed that the granules are formed into micelle structures and randomly distributed in the cytoplasm (Gerngross et al. 1994), while others proposed that the granules bud out from the plasma membrane or occur near to the cell wall at the pole of the cells (Jendrossek 2005; Pötter and Steinbüchel 2005; Hermawan and Jendrossek 2007). The mechanisms of the granule initiation, in fact, are not yet completely understood.

The time profile of P(3HB) molecular weights at 12, 24, and 48 h showed that the M_n and M_w of the homopolymer biosynthesized from jatropha oil were in the range of $2.9–4.1 \times 10^5$ Da and $9.5–16.5 \times 10^5$ Da. The range of molecular weights and the decreasing trend were commonly found in P(3HB) biosynthesized by *C. necator* (Anderson and Dawes 1990). Initially, there were no significant changes in the molecular weights of P(3HB) produced during the exponential phase

Table 5.4   Molecular weights of P(3HB) produced by *C. necator* from various carbon sources

Carbon sources	$M_n^a(\times 10^5)$	$M_w^b(\times 10^5)$	M_w/M_n^c	References
Soybean oil	3.3	12.5	3.8	Kahar et al. 2004
Olive oil	4.0	10.3	2.6	Fukui and Doi 1998
Sesame oil	2.7	5.3	2.0	Taniguchi et al. 2003

[a]Number-average molecular weight
[b]Weight-average molecular weight
[c]Polydispersity index

(12 h and 24 h). When the cells were entering the stationary phase at 48 h, the M_n and M_w of the P(3HB) produced decreased more than 37 %. The values of M_n were similar to that of P(3HB) biosynthesized by *C. necator* in limited nitrogen supply during exponential phase (about 4.8×10^5 Da) and also in a nitrogen-free media (about 2.7×10^5 Da) reported by Madden et al. (1999). In some studies, the production of P(3HB) with low molecular weights may be due to the termination reaction on growing polymer chains by chain transfer agents, which possess a hydroxyl group (Yamanaka et al. 2010). Although the 3-hydroxybutyric acid itself may act as chain transfer agent, glycerol, which was less favored by *C. necator* H16 for the P(3HB) biosynthesis and cell growth (Chee et al. 2010), was not negligible as a potential chain transfer agent. Prior to 24 h of cultivation, the lipase activity was not high. When the lipase activity became very high after 24 h, more free glycerols might be released from the triacylglycerols by the lipase. The primary or secondary hydroxyl group of the glycerols would attach to P(3HB) chains and terminate the chain propagations (Madden et al. 1999). During the production of P(3HB), the PHA depolymerase genes were found to be highly expressed in a recent study (Brigham et al. 2010). Nevertheless, the decrease in molecular weight in this study was not due to the intracellular degradation of P(3HB) as the cultivation was not prolonged to more than 48 h after the P(3HB) biosynthesis activity decreased.

Other factors that might contribute to the lower molecular weights are the bacterial strain and the extraction method used. PHA synthases possessed by the bacterial strains are the key factors determining the molecular weights and polydispersities of the P(3HB) produced (Agus et al. 2006). P(3HB) with M_n as high as 4.5×10^5 Da can be produced by the PHA synthase of *C. necator* expressed in *E. coli*. The extraction of PHA using chloroform as the solvent is a standard method which can efficiently dissolve all types of PHA. It was reported that the M_w of the P(3HB) extracted by chloroform is higher than that by sodium hypochlorite (Taniguchi et al. 2003). The extraction of PHA in this study using chloroform was shown to be an efficient method with as high as 86 % of PHA recovery. The molecular weights of the P(3HB) produced from jatropha oil was considered high and compatible with those produced from other plant oils such as soybean oil, olive oil, and sesame oil (Table 5.4) (Fukui and Doi 1998; Taniguchi et al. 2003; Kahar et al. 2004). As the M_w of the P(3HB) produced from jatropha oil was higher than 6×10^5 Da, the polymer could be used for thermoplastic applications as it is not very brittle (Braunegg et al. 1998).

Jatropha oil was found to be a feasible and excellent feedstock for P(3HB) production. P(3HB) homopolymer is known to be brittle and stiff (Sudesh et al. 2000). The incorporation of a second monomer into the homopolymer chain has been known to improve the properties of the resulting polymer (Tsuge 2002). P(3HB-co-3HV) is the first and the most studied PHA copolymer. It is more flexible and have lower melting temperature for wider processing window (Luzier 1992). Recently, Allen et al. (2010) produced P(3HB-co-3HV) with low molecular weight (1.7×10^5 Da) from saponified jatropha oil using *Pseudomonas oleovorans*. The jatropha oil had to be saponified to produce free fatty acids presumably because *P. oleovorans* was incapable of hydrolyzing the triglyceride. In the case of *C. necator* H16, the biosynthesis of P(3HB-co-3HV) with high molecular weight was achieved by using the oil directly without the need for saponification.

The best proven strategy to produce P(3HB-co-3HV) with controlled composition is by co-feeding sugars with specific precursors of 3HV such as propionic acid. Recent studies have shown that the co-feeding of 3HV precursors with triglycerides also produces P(3HB-co-3HV) with controlled composition (Lee et al. 2008). In the case of jatropha oil, it is not known whether the toxins and/or any other unknown components in the jatropha oil have any effects on the various monomer supplying pathways involved in the synthesis of the PHA copolymers. The 3HV precursors such as propionic and valeric acids have certain level of toxicity to the cells and have to be fed in a timely and controlled manner. P(3HB-co-3HV) biosynthesized from late feeding of precursors resulted in the formation of blends of copolymers having different 3HV molar fractions (Lee et al. 2008). Besides, the high content of readily accumulated P(3HB) in cells could lessen the utilization of precursors and incorporation of 3HV (Khanna and Srivastava 2007; Volova and Kalacheva 2005). However, the feeding of precursors at an early stage would have stronger toxicity effects on the cells due to lower initial cell density to tolerate the higher concentration of precursors (Shang et al. 2004). In the study by Lee et al. (2008), the cells were first grown in jatropha oil for 12 h in order to achieve adequate cell density (about 2 g/L) with P(3HB) content of 35 wt% to tolerate the toxicity effects of precursors fed later.

Lee et al. (2008) co-fed sodium valerate or sodium propionate with jatropha oil to produce P(3HB-co-3HV) with different 3HV monomer compositions. It is known that higher 3HV composition can be achieved in the copolymer produced with the feeding of substrates with smaller odd carbon numbered fatty acids such as valeric acid (C_5), propionic acid (C_3), and their salt derivatives (Akiyama et al. 1992). The substrates with smaller odd carbon numbered fatty acids are toxic to cells; thus, they are added with other cell growth- and PHA-promoting carbon sources such as sugar and oil. Jatropha oil is rich in oleic acid (42 %) which is known to improve the productivity of P(3HB-co-3HV) in *C. necator* by inducing cell growth, but at the same time, decreasing the 3HV fraction (Marangoni et al. 2000). In some studies, the 3HV fraction increased when the ratio of the precursors and main carbon sources increased (Mitomo et al. 1999; Choi et al. 2003). The composition of 3HV is generally influenced by the concentration and the ratio of the precursors. Therefore, the ratio of jatropha oil and the precursor has to be varied while fixing total amount of carbon in the culture medium.

Addition of 3HV precursors was found to significantly reduce the CDW and PHA accumulation. A 40 % reduction in CDW was observed, as compared to that obtained in the previous experiment when the jatropha oil was used as the sole carbon source. The PHA content was significantly affected when concentrations of sodium valerate exceeded 0.48 g/L. The results are in accordance with most studies (Abdelhad et al. 2009; Lee et al. 1996). Volatile short-chain fatty acids exhibit both toxic and/or inhibitive effects on the cell growth and also the PHA accumulation (Du et al. 2001b; Yu et al. 2002; Khanna and Srivastava 2007). This is because the cells need to maintain the equilibrium of pH in the cytoplasm by exporting excess protons which consume the cellular energy (Defoirdt et al. 2009). In this study, the concentration of the fatty acids at only 0.48 g/L was found to exhibit certain level of inhibition toward cell growth and PHA synthesis. Du et al. (2001b) reported that the increasing supplementation of propionic acid (3–12 g/L) inhibited the cell growth and P(3HB-co-3HV) accumulation with 15 % and 85 % decrease, respectively, in residual biomass and total PHA. Another study showed as high as 65–72 % decrease in residual biomass and total PHA when the concentration of valeric acid was increased from 2 to 10 g/L (Khanna and Srivastava 2007). Nevertheless, the salt forms of valeric acid and propionic acid used in some studies were considered to be less toxic (Loo and Sudesh 2007).

The incorporation of 3HV into the copolymer increases proportionally with the concentration of sodium valerate or sodium propionate added to the culture medium (Ishihara et al. 1996; Shang et al. 2004; Abdelhad et al. 2009). With higher concentration of precursors (3.36 and 4.32 g/L), more residual oil was present in the culture medium, indicating that the jatropha oil was less utilized. The 3HV monomer composition in P(3HB-co-3HV) produced from sodium valerate addition was higher (3–41 mol%) compared to that produced using sodium propionate (2–27 mol%) (Lee et al. 2008). The biosynthesis of P(3HB-co-3HV) from sodium valerate is more effective because the metabolic pathway is more exclusive for the biosynthesis of 3HV monomer. Sodium valerate can be converted via β-oxidation cycle into 3-hydroxyvaleryl-CoA intermediate to be incorporated directly into P(3HB-co-3HV) without catabolism (Doi et al. 1988b). A 3HV composition as high as 90 mol% had been achieved from valeric acid in some studies (Mitomo et al. 1999; Khanna and Srivastava 2007).

To further determine the effects of the mixture of jatropha oil and sodium valerate or sodium propionate on the generation of 3HV for PHA biosynthesis, sodium valerate and sodium propionate were fed by standardizing the amount of carbon present in the precursors. This enabled the calculation of conversion percentage of carbon in precursors into 3HV units. The calculation was based on the carbon content of the total 3HV units incorporated in P(3HB-co-3HV) and total carbon provided by precursors. By taking the P(3HB-co-41 mol% 3HV) copolymer produced from the mixture of jatropha oil and sodium valerate as an example, the amount of 3HV produced by the cells was 2.09 g/L. Meanwhile, the total carbon content provided to the cells by sodium valerate was 4.32 g/L. Assuming that the molar mass of the 3HV units ($C_5H_8O_2$) was 100.13 g/mol and the percentage of carbon in the 3HV unit was 60 %, the carbon content of the total 3HV produced (2.09 g/L) was 1.25 g/L. This means that the conversion percentage of carbon in sodium valerate into 3HV units was about 29 %. In general, the conversion percentage of sodium valerate into 3HV units in the presence of jatropha oil was in the range of 24–31 % with an average of 28 %, which is 1.6-fold higher compared to the average conversion percentage using sodium propionate (17 %). The 3HV molar fraction produced from sodium valerate is almost twofold of that from sodium propionate (Bhubalan et al. 2008). The low conversion percentage of short-chain fatty acids is common and may be due to the accumulation of undissociated fatty acids in the cells which inhibit the substrates utilization (Wang et al. 2010). Nevertheless, the conversion percentages of sodium valerate and sodium propionate into 3HV monomer in the presence of vegetable oils are relatively higher than those reported for other concentrations of carbon sources (Doi et al. 1987a, 1988b; Lee et al. 1996, 2008; Du et al. 2001b; Song et al. 2001; Bhubalan 2010).

The incorporation of 3HV has been known to improve the thermal and mechanical properties of P(3HB) (Savenkova et al. 2000; Volova and Kalacheva 2005; Zakaria et al. 2010). The melting point of P(3HB) was reduced from 183 °C when co-monomer was incorporated into P(3HB) homopolymer. Generally, T_m decreased from 150 °C to a minimum of 131 °C when the 3HV molar fraction was at 22 mol%, but increased again to 162 °C when the 3HV molar fraction was 42 mol%. The trend of T_m with a decrease followed by an increase showed a pseudoeutectic melting behavior of the isodimorphism in the P(3HB-co-3HV) with the transition of crystal phases from P(3HB) lattice to P(3HB-co-3HV) lattice (Bluhm et al. 1986; Mitomo et al. 1993; Yoshie et al. 2001). According to several studies, the pseudoeutectic composition for the transition of crystal phase ranged approximately from 30 to 56 mol % of 3HV (Bluhm et al. 1986; Mitomo et al. 1993; Yoshie et al. 2001). A lower pseudoeutectic composition at 25 mol% of 3HV and narrow changes of T_m for the second heating (154–174 °C) was reported (Keenan et al. 2006).

Two separate melting peaks were observed from the second heating endothermic curve of P(3HB-co-3HV) with 4, 10, and 22 mol% of 3HV. Multiple melting peaks are commonly reported for P(3HB-co-3HV) copolymer (Mitomo et al. 1995; Yoshie et al. 1995; Wang et al. 2001; Žagar et al. 2006). The multiple peaks may indicate the coexistence of two copolymers with different compositions of 3HV monomer (Mitomo et al. 1995). The existence of the multiple peaks may also be caused by the rearrangement of crystal morphology during the crystallization process. Abe et al. (1999) reported that during the preparation of solution-cast film, the polymers with different 3HV content might rearrange into lamellar crystals with different thickness due to the variation in crystallization rate. Another rearrangement of the crystal structure also occurred during the heating of DSC by the melting-recrystallize-remelting process to form well-organized lamellae from the cooled unordered lamellae caused by the incorporation of 3HV monomers, which act as a disruptor (Yoshie et al. 1995; Saito et al. 2001).

The low ΔH_m obtained in several studies indicated low crystallinity of the P(3HB-co-3HV) copolymer produced, which remained in the amorphous phase with no cocrystallization (Da Silva et al. 2005; Feng et al. 2002). Quenched P(3HB-co-3HV) with intermediate compositions (34–71 mol%) did not crystallize during the second heating (Scandola et al. 1990). This explains why no melting peaks at lower temperatures were observed for P(3HB-co-3HV) with 3HV content as high as 27 and 42 mol%. Since the crystallinity is so low, the higher melting temperature might not reflect the true phase change of crystal but might indicate the coexistence of copolymers with a lower 3HV fractions or rearrangement of crystal structure during the heating of DSC. As with other studies, increased incorporation of 3HV or 3HHx monomer into the copolymers decreased the T_g (Savenkova et al. 2000; Keenan et al. 2004). All the thermal degradation temperatures of P(3HB-co-3HV) at 5 % weight loss were similar to that of P(3HV) homopolymer at 258 °C (Shen et al. 2009) and much higher than their melting temperatures. Therefore, the incorporation of 3HV monomer improves the thermal properties of P(3HB) for broader thermal processing window (Bengtsson et al. 2010).

The M_w of P(3HB-co-3HV) produced from the mixture of jatropha oil and precursors by *C. necator* H16 were moderately high, in the range of 9.0–18.4 × 10^5 Da with polydispersity of 3.1–3.9. Molecular weight is a crucial factor for improving the physical, mechanical, and thermal properties of the copolymers but the effects are complicated. For P(3HB-co-3HV) with the same 3HV composition but increasing molecular weight, the lamellar thickness, amorphous layer thickness, melting temperature and tensile strength increased, while the melting enthalpy, crystallinity, and crystallization temperature of second scans were found to decrease (Luo et al. 2002). Nevertheless, increasing 3HV composition did not show any variation on the M_w of the copolymer produced (Volova and Kalacheva 2005).

The thermal properties of moderately high M_w P(3HB-co-3HV) produced from jatropha oil and precursor were comparable to those produced from other plant oils (Lee et al. 2008). Jatropha oil was found compatible to be used in combination with precursor carbon sources for the biosynthesis of PHA copolymers by *C. necator* H16. The compositions of the P(3HB-co-3HV) copolymers were controllable by varying the concentration of the 3HV precursors. Jatropha oil supported both good cell growth and the biosynthesis of PHA copolymers by *C. necator* H16. In conclusion, Jatropha oil is a potential feedstock for the large-scale production of various types of PHAs.

Chapter 6
Potential Applications of PHA

Abstract Polyhydroxyalkanoate (PHA) is an attractive material because it can be produced from renewable resources and because of its plastic-like properties. In addition, PHA can be degraded by the action of microbial enzymes. Although PHA resembles some commodity plastics, the performance and cost of PHA are not yet good enough for widespread applications as plastic materials. Therefore, the PHA commercialization attempts by many industries for bulk applications have been challenging. However, PHA also possesses interesting properties that can be developed for non-plastic applications. This chapter describes some new niche applications for PHA in cosmetics and wastewater treatment.

Keywords Batik dye • Cosmetics • Electrospinning • Facial oil • Toxicity studies • Oil blot • Polyhydroxyalkanoate (PHA) • Polyhydroxybutyrate (PHB)

PHA has a wide array of useful properties and potential applications (Orts et al. 2008). The potential applications of PHA include packaging, disposable items, and for medical use (Khanna and Srivastava 2005a), covering areas such as medicine, agriculture, tissue engineering, nanocomposites, and polymer blends (Philip et al. 2007). PHA can also be used as stereoregular compounds which can serve as chiral precursors for the chemical synthesis of optically active compounds (Oeding 1973; Senior and Dawes 1973). PHA that contains 4HB monomer such as P(3HB-*co*-4HB) is suitable for biodegradable drug carriers (Turesin 2001). Such compounds are particularly useful as biodegradable carriers for long-term delivery of drugs, medicines, hormones, antibiotics, insecticides, and herbicides (Khanna and Srivasta 2005a; Chen and Wu 2005). PHA is also an interesting class of biomaterials for the development of tissue-engineered cardiovascular products (Williams and Martin 2002).

To mimic the real microenvironment of extracellular matrix for cell growth, the scaffold should provide a highly biocompatible 3D biodegradable polymer substrate to enable cell adhesion, migration, proliferation, and differentiation function to develop into a tissue-like structure (Deng et al. 2002). P(3HB-*co*-3HHx) is less porous when the 3HHx component is increased, making it smoother (Wang et al. 2005b). Fibroblast cells are reported to attach to smoother surface

(Wang et al. 2005b). The material surface properties affect the initial cellular events on the cell-material interface (Wang et al. 2005b). It is therefore desirable to modify the material surface to suit the intended application, without altering other properties of the scaffold, such as mechanical strength or thermal properties (Williams and Martin 1999; Liu et al. 1999).

Crystallinity at polymer surface is also responsible for difference in surface morphology between P(3HB) and P(3HB-co-3HHx) (Mori et al. 2008). Effective surface modifications include changes in chemical group functionality, surface charge, hydrophilicity, hydrophobicity, and wettability (Williams et al. 1999; Liu et al. 1999; Sacristan et al. 2000). Modification of a polymer surface can be achieved by various chemical or physical processes including plasma-ion beam treatment, electric discharge, surface grafting, chemical reaction, vapor deposition of metals, and flame treatment (Williams et al. 1999). Hydroxyapatite that was blended into P(3HB) increases elasticity and stress tolerance (Wang et al. 2005a). Protein has both hydrophobic and hydrophilic regions causing a selective attachment of protein to a surface, and the area with appropriate proportions of hydrophobicity and hydrophilicity will be favored for protein attachments (Wang et al. 2003). Plasma treatment improves surface polarity, but the incorporations of polar components decrease the hydrophobicity, and changes the surface chemical state and electrical charges (Qu et al. 2005).

The surface properties of a biomaterial, especially hydrophilicity, influence cell adhesion to the materials (Chanvel-Lesrat et al. 1999; Qing et al.1999; Zhao and Geuskens 1999; Furukawa et al. 2000). Cells attach to the biomaterial by filopodia, forming bridging cells for the attachment of other cells to form cell aggregates (Deng et al. 2002). Hydrolysis process by lipases or sodium hydroxide will generate more hydroxyl groups that will lead to increased hydrophilicity, improving the capacity for cells to adhere to the polymer surfaces (Yang et al. 2002). The higher the hydrophilicity of a material surface, the stronger the cells attach to that material (Yang et al. 2002). Higher protein absorption and protein conformational changes occur on hydrophobic surface because it is difficult for proteins to deabsorb from hydrophobic surface due to hydrophobic bonds (Dee et al. 2002). P(3HB-co-3HHx) containing high 3HHx content tends to be more hydrophobic (Qu et al. 2006).

On the other hand, PHA powder (Lawrence et al. 2005) treated with UV radiation increased its hydrophilic functional groups (increase in polar groups C–O and C=O), degradation rate, and mechanical property (Shangguan et al. 2006). Although treating polymers with chemicals is proved to improve the polymer properties, it is better to minimize its usage because the residual chemicals may exhibit side effects if they are not removed thoroughly. Mechanical properties of P(3HB) can be improved when blended with P(3HHx) (Zhao et al. 2003). The high crystallization degree and rapid crystallization rate of P(3HB) create pores and protrusion on the film surface (a coralloid surface) that might prohibit the attachment and growth of mammalian cells (Zhao et al. 2003). The P(3HHx) in the P(3HB)/P(3HHx) blend reduced the crystallization degree and crystallization rate (Zhao et al. 2003).

In blend polymers, P(3HB) crystallizes to form the crystalline domains that act as physical crosslinkers and fillers, and P(3HHx) forms the amorphous domain (Zhao et al. 2003). It was found that mouse fibroblast cell line L929 grew better in the blend of P(3HB) and P(3HHx) films rather than on the individual films, but viable cells on P(3HHx) alone were 216 times more than that of P(3HB) (Yang et al. 2002). These results showed that P(3HHx) promotes cell growth, an important criteria for biomaterial. Lipase was shown not to be able to degrade P(3HHx) efficiently or the degraded product shows mixed effects, showing little change on its surface pore size (Yang et al. 2002). At a ratio of 1:1 for P(3HB)/P(3HHx) blends, the non-dispersion component of surface free energy is maximum, leading to a maximal total surface free energy (Zheng et al. 2005). The higher the surface free energy of a blend film, the more amount of protein will be absorbed to the film (Zheng et al. 2005). That was the reason protein absorption and rabbit articular cartilage chondro-cytes adhesion were detected on the blended films (Zheng et al. 2005).

6.1 Some New Applications of PHA

The biocompatibility and biodegradable plastic-like properties of PHA have evoked its potential use in several applications such as medical, tissue engineering, packaging industry as well as cosmetics and skin care industry (Mauclaire et al. 2010; Sudesh et al. 2007; Valappil et al. 2007). PHA cast films were also investi-gated as potential facial oil blotting material (Sudesh et al. 2007). Three types of PHA cast films namely P(3HB), P(3HB-co-3HV), and P(3HB-co-3HHx) copoly-mer were tested for oil absorbability, retention, and oil-indication properties. It was reported that all the tested PHA films revealed similar oil-absorption char-acteristics. The PHA films tested were able to absorb sebum on the skin even with-out the addition of lipophilic additives such as mineral oil and zinc stearate, which are usually added into commercial facial oil blotting films. The oil absorbed onto the films was indicated by greater transparency and the changes were more obvi-ous on P(3HB-co-3HHx) films. The oil retention of P(3HB-co-3HHx) film was found to be around 80 %. The PHA films were able to absorb oil efficiently when reused after being washed with detergent. Commercial facial oil blotting films lose their ability to absorb oil upon washing with detergent probably because of the removal of oil absorbents added to the film surface. This showed the reusability of PHA film for this application.

6.2 Biomedical Applications of PHA

As PHAs are biosynthetically produced, they may have several advantages for bio-medical applications. Without the usage of chemical catalysts, toxicological prob-lems can be avoided. PHAs are particularly attractive because they are bioabsorbable and biocompatible. The metabolism and excretion of some monomers incorporated into PHAs are well understood. For example, the monomeric component of P(3HB) (R)-3-hydroxybutyric acid (3HB), is a ketone body present at concentrations of 3–10 mg per 100 mL blood in healthy adults (Williams and Martin 2002; Hocking and Marchessault 1994). The monomeric component of P(4HB), 4-hydroxybutyric acid (4HB), can also be found widely distributed in the brain, kidney, heart, liver, lung, and muscle of the mammalian body (Nelson et al. 1981). The P(3HB) shows high crystallinity with a melting temperature (T_m) of 175 °C and a glass transition temperature (T_g) of 0–4 °C. Therefore, due to its high brittleness, poor processability, and slow degradation, P(3HB) is thought to be of limited use (Iwata et al. 2003). Nevertheless, studies in this field of interest have shown great progress over the past 20 years, and it is now possible to design and synthesize various kinds of PHA to overcome the inferior properties of P(3HB) (Sudesh 2004). As such, the number of PHAs currently under evaluation as biomaterial has expanded to five, that is P(3HB), P(4HB), poly[(R)-3-hydroxybutyrate-co-4-hydroxybutyrate] [P(3HB-co-4HB)], poly[(R)-3-hydroxybutyrate-co-(R)-3-hydroxyvalerate] [P(3HB-co-3HV)], and poly[(R)-3-hydroxybutyrate-co-(R)-3-hydroxyhexanoate] [P(3HB-co-3HHx)]. To date, PHA and its composites are thought to have good potentials as emerging materials for medical devices such as sutures, bone plates, surgical mesh, and cardiovascular patches (Chen and Wu 2005).

6.3 Electrospun PHA Tissue-Engineering Scaffolds

Another promising potential of PHA in medical field is for the fabrication of scaffold as an ideal structure that can replace the natural extracellular matrix (ECM) until the host cells can repopulate and resynthesize a new natural matrix.

Scaffolds should be well integrated in the tissue of the host without eliciting adverse immune response. The regeneration of specific tissues aided by biomaterials has been shown to be dependent on the porosity and pore size of the supporting 3D structure (Cima et al. 1991). A large surface area promotes cell attachment and proliferation, whereas a large pore volume is necessary to accommodate and deliver sufficient cell mass needed for tissue repair. Porosity and interconnectivity are also essential for diffusion of nutrients and gases and for the removal of metabolic wastes resulting from activity of the cells. The scaffold should have sufficient mechanical strength to retain its structure after implantation and its degradation rate must be tuned in accordance to the growth rate of the cells, in such a way that by the time the injury site is regenerated, the scaffold is completely degraded. Bulk erosion is undesirable because it will cause the scaffolds to fail prematurely. New polymer processing approaches are in demand to create degradable porous scaffolds that can support the hierarchical structures of many tissues ranging between 0.1 and 1.0 mm (Griffith 2000). Electrospinning has emerged as one of such methods offering simplicity and versatility in preparing such biomaterials (Cima and Cima 1996).

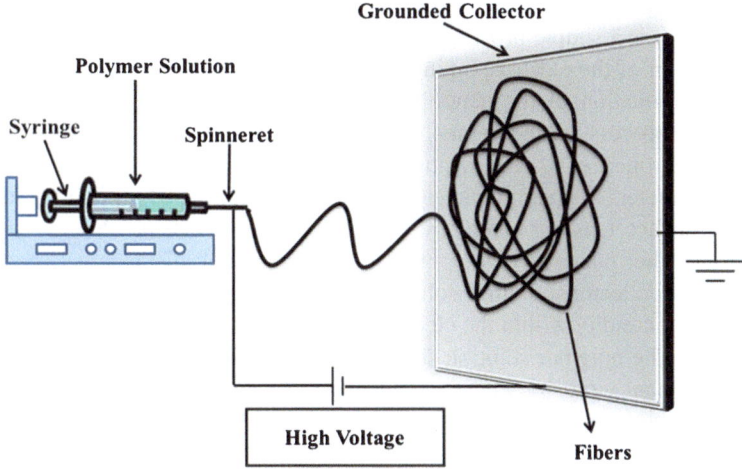

Fig. 6.1 Schematics of the horizontal electrospinning experimental set up

Electrospinning is an electrostatic spinning process capable of assembling fibrous polymer mats consisting of fibers in the range of several microns down to less than 100 nm (Frenot and Chronakis 2003; Tan and Obendorf 2007). Electrospinning technique was first patented by Formhals in 1934 (Zong et al. 2005). In the 1960s, Taylor documented that the shape of polymer droplet at the spinneret tip was a cone and that jets were ejected from the vertices of this cone when an electric field was applied (Subbiah et al. 2005). This conical shape was later referred to in the literature as the 'Taylor Cone'. The basic experimental set up for electrospinning is presented in Fig. 6.1. Electrospinning can be performed using either a horizontal or vertical setup. The horizontal set up is more advantageous for the less viscous polymer spinning solutions. In the electrospinning process, a polymer solution is charged with high voltage to create repulsive force within the polymer solution. The repulsive force then overcomes the surface tension of the solution at a critical voltage causing the eruption of a jet from the tip of 'Taylor Cone'. The ejected jet is subjected to bending instability and further splitting due to whipping process to form thinner fiber jets (Deitzel et al. 2001).

As the polymer solvent evaporates, a charged polymer fiber forms on the collector which bears electrical charges of opposite polarity (Zhong et al. 2007). The fibers that are produced by electrospinning are electrically charged, and thus these fibers can be guided by electrical fields (Huang et al. 2003). Fabricated electrospun polymer nanofibers have small pores and high surface area to volume ratio due to its small fiber diameter as compared to commercial nonwoven fabrics. Hence, electrospun nanofibers have found vast applications in optical and chemosensor materials, nanofibers with specific surface chemistry for tissues scaffolds, nanocomposite materials, drug delivery systems, wound dressing, filtration membranes, and protective clothings (Gupta et al. 2005b; Azad 2006). Although conceptually it is a simple process, electrospinning has significant challenges. The major criticism of the electrospinning technique has been the comparatively slow, batchwise production of the nanofibers. However, increasing commercialization efforts of the process has led to much technical progress in recent years.

The final properties of electrospun fibers are affected by numerous system and process parameters. System parameters include the polymer M_w, polymer solution viscosity, surface tension, conductivity, and dielectric constant. Process parameters, on the other hand, involve applied electric potential, distance between spinneret tip and collector, polymer solution flow rate, and also the temperature, air velocity, and humidity within the electrospinning chamber (Demir et al. 2002; Ren et al. 2008). The polymer solution should possess a concentration high enough to achieve sufficient viscosity where polymer chain entanglements occur, but not too high that the viscosity prevents polymer motion under an electrical field. The surface tension of the electrospinning solution should be low enough while charge density and viscosity are held high enough to avoid the jet from breaking up into droplets before complete solvent evaporation. By changing the distance between syringe needle and grounded collector, it could alter the intensity of the electrical field which results in morphological changes in the electrospun fibers. An increase in the applied electric field is known to decrease the bead density in fibers regardless of the polymer concentration. Higher flow rate of polymer solution is often associated with larger fiber diameter (Li and Xia 2004; Cheng et al. 2008).

Nanofiber matrices with different fiber diameters exhibit a wide range of surface properties, superior mechanical properties, and porosity compared with other forms of the material. These properties impart nanofiber matrices with structural and functional similarity to the native ECM structure. Thus, many studies are exploring the use of nanofiber matrices as scaffolds for tissue-engineering and regenerative medicine. Numerous polymer biomaterials have been used for scaffold fabrication, including non-biodegradable and biodegradable polymers, with the latter consisting of natural and synthetic polymers. Biodegradable polymers are generally favored over non-biodegradable polymers because the long-lasting nature of the non-biodegradable polymers can impede tissue renewal and remodeling. Polymer biodegradation via the combined effect of enzymatic and hydrolytic activities, generates space within the scaffold that facilitates perfusion and deposition of newly synthesized ECM. So far, over a hundred different biodegradable polymers have been successfully electrospun and more than a third of these have been applied for various tissue-engineering purposes. A summary of polymers including PHA used in electrospinning, their potential applications, and evaluation criteria used in each study reported between the year 2000 and 2008 is given in Table 6.1.

Electrospinning of PHA is still relatively new in scaffold fabrication. To date, P(3HB) and P(3HB-co-3HV) are the most common microbial polyesters to be electrospun into tissue-engineering scaffolds. Suwantong et al. (2007) prepared ultrafine electrospun fiber mats of P(3HB) and P(3HB-co-3HV) as scaffolding materials for skin and nerve generation. In their study, they evaluated the in vitro biocompatibility of these fibers using mouse fibroblasts and Schwann cells

Table 6.1 Summary of polymers used in electrospinning studies

Study	Biomaterial	Applications		Characterization	Ref.
1	PHB, PHBV	Nonwoven scaffold	TE	In vitro Schwann cell culture, cytotoxic evaluation, 3-(4.5-dimethylthiazol-2-yl) (MTT) assay	(Suwantong et al. 2007)
2	PHBV, hydroxyapatite (HA)/PHBV	Nonwoven, aligned scaffold	TE	(SEM)	(Tong and Wang 2007)
3	PHBV, PHBV-PLLA, PHBV-PLGA	Nonwoven scaffold	TE	SEM, oxygen plasma treatment, fluorescence microscopy, MTT assay	(Ndreu et al. 2008)
4	Gelatin-containing PHBV	Biomedical application		Attenuated total reflection-Fourier transform infrared spectroscopy (ATR-FTIR), electron spectroscopy for chemical analysis (ESCA), atomic force microscopy (AFM)	(Meng et al. 2008)
5	PHBHHx	Nonwoven scaffold	TE	In vitro hMSCs culture, fluorescence stain, MTT assay, alizarin red S stain	(Yu et al. 2008)
6	PHB/PHBHHx PHB/P3HB4HB	Nonwoven scaffold	TE	SEM, Wide Angle X-ray Diffraction (WAXD), density test, mechanical test, assay of cell attachment and viability	(Li et al. 2008)
7	PLLA	Nonwoven scaffold	TE	SEM, mercury porosimeter, AFM, contact angle test, in vitro neural stem cell culture	(Yang et al. 2004)
8	PLLA	Nonwoven scaffold	TE	SEM, AFM, in vitro endothelial cells culture, assay of cell attachment and viability	(Xu et al. 2004)
9	PGA, PDLLA, PLLA, PLGA, PCL	Nonwoven scaffold	TE	SEM, tensile test, degradation evaluation, in vitro chondrocytes and hMSCs culture, cell proliferation analysis	(Li et al. 2006)
10	PLLA	Nonwoven, aligned scaffold	TE	SEM, in vitro mouse cerebellum stem cell culture	(Yang et al. 2005)
11	PLLA	Nonwoven, aligned scaffold	TE	In vitro primary motor neurons culture, immunocytochemistry	(Corey et al. 2008)
12	PLLA, PDLLA, Poly(ethylene glycol)(PEG)-PLLA, PEG-PDLLA	Nonwoven scaffold	TE	Contact angles measurement, SEM, in vitro osteoprogenitor cell culture, alkaline phosphatase activity	(Badami et al. 2006)
13	PLLA, Polystrene	Nonwoven aligned and random scaffold	TE	In vitro human dermal fibroblasts culture, MTT assay	(Sun et al. 2007)

(continued)

Table 6.1 (continued)

Study	Biomaterial	Applications		Characterization	Ref.
14	PLLA, PLGA, PGA	Nonwoven scaffold	TE	SEM, ESCA, in vitro fibroblast culture, cell adhesion, and proliferation assay	(Park et al. 2007)
15	PCL	Nonwoven scaffold	TE	SEM, in vitro human coronary smooth muscle cell culture	(Venugopal et al. 2005)
16	PCL	Nonwoven scaffold	TE	SEM, TEM, in vitro human dermal fibroblast culture	(Zhang et al. 2005)
17	PCL	Nonwoven scaffold	TE	SEM, in vivo rat model	(Shin et al. 2004)
18	PGA	Nonwoven scaffold	TE	SEM, TEM, in vitro rat cardiac fibroblast culture, in vivo rat model	(Boland et al. 2004)
19	Elastin-mimetic	Nonwoven scaffold	TE	SEM	(Nagapudi et al. 2005)
20	Silk fibroin, silk/poly(ethylene oxide) (PEO)	Nonwoven scaffold	TE	SEM, FTIR, X-ray photoelectron spectroscopy (XPS)	(Jin et al. 2002)
21	Silk, silk/PEO	Nonwoven scaffold	TE	SEM, XPS, Differential Scanning Calorimetry (DSC), mechanical evaluation, in vitro human bone marrow stromal cell culture	(Jin et al. 2004)
22	Silk	Biomedical applications	TE	SEM, TEM, WAXD	(Zarkoob et al. 2004)
23	Silk fibroin	Nonwoven scaffold for wound healing	TE	SEM, porosimetry, in vitro human keratinocyte and fibroblast cultures	(Min et al. 2004a)
24	Silk/chitosan	Nonwoven scaffold for wound dressing	TE	SEM, viscosity analysis, conductivity	(Park et al. 2004)
25	Chitin, chitosan	Wound dressings		SEM, ^{1}H Nuclear magnetic resonance (NMR), FTIR, WAXD, DSC, thermogravimetric analysis (TGA)	(Min et al. 2004b)
26	Chitosan	Nonwoven, aligned scaffold for cartilage	TE	SEM, mechanical evaluation, degradation, in vitro canine chondrocyte culture	(Subramanian et al. 2004)
27	Chitosan	Nonwoven biomaterial		SEM	(Geng et al. 2005)
28	Chitosan/PEO	Nonwoven scaffold, drug delivery	TE	SEM, XPS, FTIR, DSC	(Duan et al. 2004)
29	Gelatin	Nonwoven scaffold	TE	SEM, mechanical evaluation	(Huang et al. 2004)

as reference cell lines and compared the performance with solution-cast films. While the fibroblasts were found to adhere better to the fibrous scaffolds, the Schwann cells preferred the flat surfaces of films. Tong and Wang carried out the electrospinning of aligned P(3HB-co-3HV) fibers, which was incorporated with 20 % carbonated hydroxyapatite (HA) nanospheres (Tong and Wang 2007). They reported that a nearly perfect alignment of fibers was possible with a high drum rotating speed of 3000 rpm. Ndreu et al. (2008) performed the electrospinning of P(3HB-co-3HV) and its blends with PLLA, PLGA, and P(L,DL)LA. In addition, they also performed in vitro studies using human osteosarcoma cells and found out that among the scaffolds, the blend of P(3HB-co-3HV) and PLLA showed the best results in cell proliferation and attachment. In a more recent study, Meng et al. (2008) fabricated a nanofibrous scaffold by co-electrospinning P(3HB-co-3HV) and gelatin at a ratio of 50:50. When fibroblasts were cultured on this scaffold, the number of cell growth was much higher than on P(3HB-co-3HV) scaffold. Besides P(3HB) and P(3HB-co-3HV), P(3HB-co-3HHx) is also another potential material for electrospun scaffold. Yu et al. (2008) prepared P(3HB-co-3HHx) scaffolds of different surface characteristics using compression molding, solvent casting, and electrospinning methods. In their study of in vitro cultivation of human MSCs, they found that the electrospun P(3HB-co-3HHx) scaffold outperformed the other scaffolds in terms of cell proliferation and viability as well as inducing the MSCs to differentiate into bone-like cells.

6.4   PHA-Based Nanocomposite Materials for Textile Dye Wastewater Treatment

Besides the fabrication of PHA films for oil blotting application, the hydrophobic property of PHA has also led to the use of this material for dye removal via adsorption in textile wastewater treatment. The potential application of PHA films as facial oil adsorbing material suggested that PHA films can also be used to adsorb other hydrophobic compounds. Many textile dyes are hydrophobic and may readily adsorb onto PHA films. Therefore, PHA films maybe used to remove textile dyes from wastewater. Solvent-cast P(3HB) films were found to remove approximately 38 % of color from textile dye wastewater. Electrospun PHA films may show better ability in adsorbing hydrophobic textile dyes.

6.4.1   Effect of CHCl₃ Neat Solvent on the Electrospinnability of P(3HB)

Initial experiments to determine the suitable electrospinning parameters were carried out using P(3HB) alone in CHCl₃ as the sole solvent. The morphology of films electrospun from different solution concentrations are shown in Fig. 6.2.

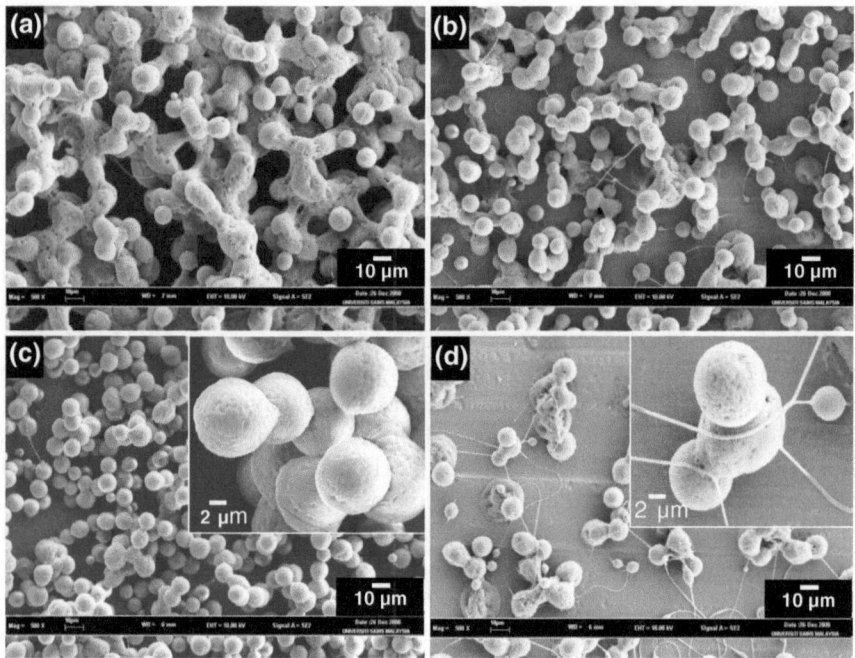

Fig. 6.2 SEM micrographs of **a** 1 % (w/v); **b** 2 % (w/v); **c** 3 % (w/v), and **d** 4 % (w/v) P(3HB) in CHCl₃ electrospun at applied voltage and extrusion rate of 15 kV and 40 μL min⁻¹ respectively. Inset: magnified views

At concentrations below 4 % (w/v), electrospinnability of P(3HB) was not achieved as there was insufficient polymer chain overlap in the solution (Gupta et al. 2005b). Structures consisting only of beads were observed and the degree of coalescence between the beads reduced with increasing solution concentration. The beads-only structure indicated that the polymer chain entanglement was almost negligible at concentrations lower than 4 % (w/v). Bead formation (also known as electrospraying) is a result of Rayleigh instability which causes the jet to breakdown into droplets (Shenoy et al. 2005a). According to Deitzel et al. (2001), the average size of electrosprayed droplet is associated with the shape of jet originating surface. At 4 % (w/v), beads interconnected by incipient fibers were observed. It marked the onset of certain degree of chain entanglement in the polymer solution (Eda and Shivkumar 2007). This enabled the jet between polymer droplets to form nanofibers instead of complete jet breakup (Gupta et al. 2005b). Fiber formation is often achieved at high concentration when sufficient chain entanglement overcomes the electrically driven jet and allows further elongation of the jet by electrostatic stress (Cheng et al. 2008). Thus, to further improve the cohesion force it was necessary to increase the solution concentration to 5 % (w/v). However, the high solution viscosity prohibited continuous flow of the polymer solution to the capillary tip and suppressed electrospinning. Such constraint was also reported in previous studies using polymers such as polyethylene oxide (PEO) and poly-D-lactide (PDLA) (Subbiah et al. 2005). Beads-free nanofibrous structures were not obtained even with all the applicable voltages and extrusion rates. These findings indicated that the balance between polymeric chain entanglement and electrical force could not be reached in the P(3HB)-CHCl₃ system.

6.4.2 Effect of CHCl₃/DMF Mixed Solvent on the Electrospinnability of P(3HB)

According to the literature, addition of a co-solvent can significantly lower the critical concentration needed for fiber initiation. An appropriate co-solvent can increase the electrical energy of the polymer–solvent system which is favorable for fiber formation (Shenoy et al. 2005a) and reduces the dependence on high solution concentration. The addition of dimethylformamide (DMF) as a co-solvent dramatically improved the electrospinnability of 4 wt% (w/v) P(3HB). The presence of DMF, a marginal solvent for P(3HB) could have increased the cohesion force between P(3HB) polymer chains. Similarly, it was reported that the addition of dichloromethane (DCM), a poor solvent for PLLA increased the polymer chain entanglement in dioxane (Cheng et al. 2008). In CHCl₃/DMF system, 4 wt% (w/v) P(3HB) tend to form thermoreversible physical gelation upon cooling to room temperature. This, indirectly indicates increase in polymer solution viscosity. Low affinity solvents tend to result in increased polymer–polymer interactions which lead to higher chain aggregation and rapid gelation. This is attributed to liquid–liquid and solid–liquid phase transition (Shenoy et al. 2005b).

Owing to the poor conductivity of P(3HB)-CHCl₃ system, nanofiber formation was not possible. The presence of DMF molecules enhanced the conductivity of the solvent system due to its high dielectric constant (Wang et al. 2008). This resulted in higher net charge density of the solution and thus more electrical charges were carried in the electrospinning jet. Thus, the stretching force of the jet was enhanced resulting in fiber formation (Shenoy et al. 2005a; Patra et al. 2009). Figure 6.3 illustrates the morphology of P(3HB) fibers electrospun from CHCl₃/DMF system. When CHCl₃/DMF at a ratio of 8:2 was used to dissolve P(3HB), fiber onset was achieved at an applied voltage as low as 10 kV. However, the resultant fibers had spindle-like bead defects (Fig. 6.3a). As the applied voltage was further increased to 15 kV, smooth homogeneous fibers were obtained (Fig. 6.3c). The boiling point of CHCl₃ and DMF is 61.2 and 153 °C, respectively. Hence, the addition of DMF lowers the evaporation rate of the mixed solvent which promoted formation of smooth surface and extended stretching time of the fibers (Cheng et al. 2008). The average diameter of the electrospun nanofibers was 450 nm. Higher applied voltage imposed greater extentional force on the extruded jet, and thus caused the transition from spindle-like beads to smooth surfaced nanofibers (Eda and Shivkumar 2007). Similar morphology transition was also observed in polystyrene fibers electrospun at 15 kV from THF/DMF solvent at 50/50 ratio (Lee et al. 2003). As the applied

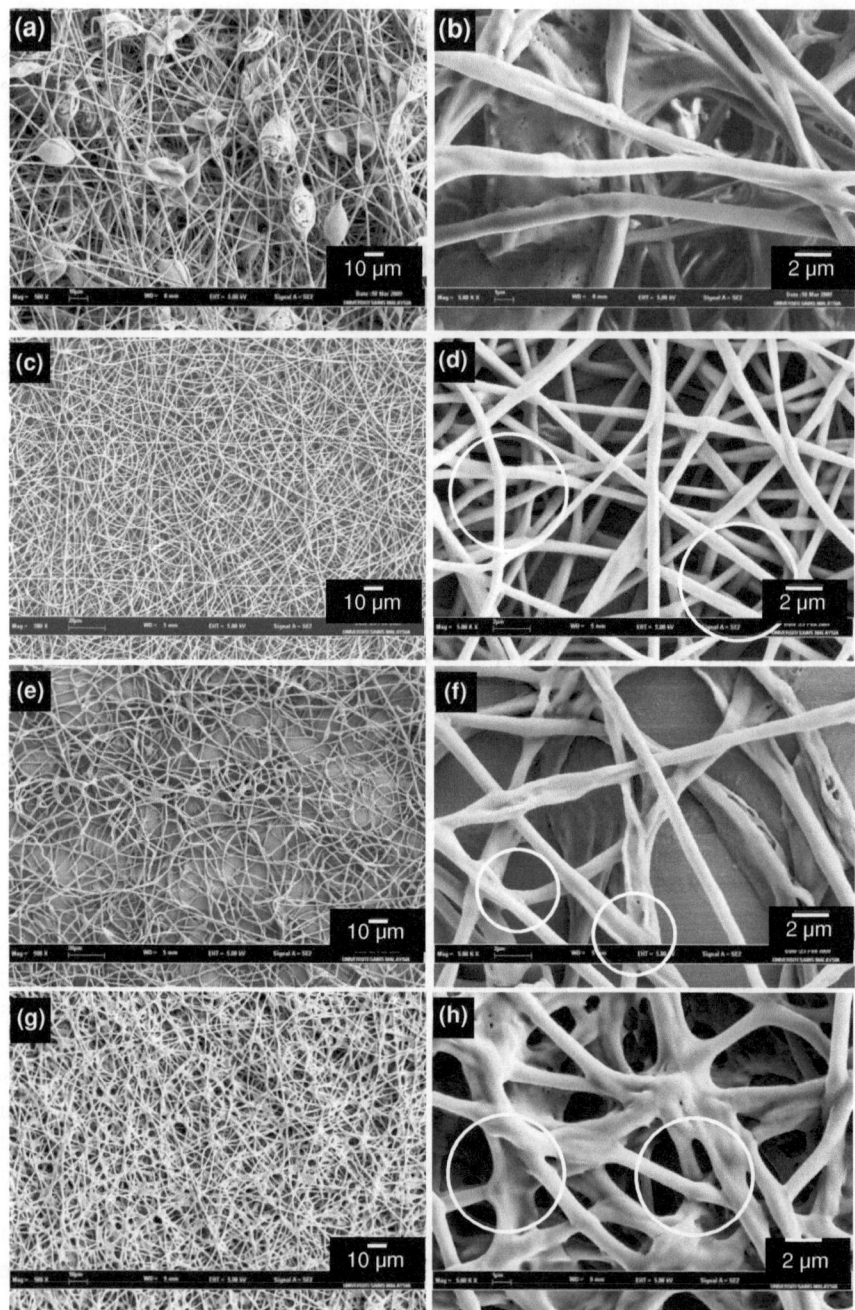

Fig. 6.3 Morphological features of 4 % (w/v) P(3HB) in CHCl₃/DMF mixed solvent (ratio:8:2) electrospun at **a** & **b** 10 kV; **c** & **d** 15 kV, and **e** & **f** 30 kV with an extrusion rate of 40 μL min⁻¹. SEM micrographs **g** & **h** are of similar polymer solution electrospun at applied voltage and extrusion rate of 15 kV and 60 μL min⁻¹, respectively. SEM micrographs **b, d, f,** & **h** are magnifications of **a, c, e,** and **g,** respectively

Fig. 6.4 Electrospun P(3HB) nanofibrous film without TiO$_2$ **a** before and **b** after immersion in batik dye waste water under solar illumination of approximately 98,000 Lux for 3 h. The batik dye waste water **C** before and **D** after treatment with electrospun P(3HB) nanofibrous film revealed approximately 80 % of color removal through adsorption of dye molecules on the film and self-photolysis of the Batik dye solution

voltage was increased to 30 kV, ridges and surface undulations formed on the electrospun fibers (Fig. 6.3f). This might be due to incomplete splaying of the jet as it was extruded more rapidly (Eda and Shivkumar 2007). There was also lack of uniformity in the morphology of the fibers. On the other hand, as the extrusion rate was increased from 40 to 60 μL min^{-1}, fibers were seen to coalesce at their contact points and some of the fused fibers were also observed (Fig. 6.3h). This suggests that evaporation of the solvent was incomplete when the fibers were stacked upon one another during electrospinning. This might be due to the reduction of the jet traveling time with increasing voltage (Jalili et al. 2005). According to Ishii et al. (2007), coalescence of nanofibers tends to increase the durability of nanofibers against breakage. However, it should be noted that such coalescence also reduces the surface area of electrospun fibers.

Upon optimizing the electrospinning, the nanofibrous film (Fig. 6.3d) was tested for its ability to adsorp dyes from textile wastewater. For this purpose, wastewater from the Batik industry was used. Batik is a traditional form of art practiced in several countries of Southeast Asia (Ali and Suhaimi, 2009). This industry is usually operated in small scale as backyard or cottage industries with many of its workshops or factories located adjacent to coastline like those found in the Malaysian states of Penang, Kelantan, and Terengganu. Due to the lack of stringent or regulated Batik waste disposal system, the Batik effluent is usually partially diluted and released into nearby rivers. Various methods are being studied for treating this Batik dye effluent. One of this method is the utilization of PHA nanofibrous film. It was found that almost 80 % of color was removed from the Batik industry wastewater by adsorption and self-photolysis without photocatalysis (Fig. 6.4). However, adsorption is merely phase transference of the dye pollutant (Zhou et al. 2008; Gupta and Suhas 2009). The problem is not solved completely. Thus, P(3HB) nanofibers were incorporated with TiO_2 (P-25, ca. 80 % anatase, 20 % rutile; BET area, ca. 50 m^2g^{-1}) nanoparticles to develop a novel wastewater treatment method combining both adsorption and advanced oxidation process (Kuo and Lin 2009; Kansal et al. 2009; Khataee et al. 2009; Khataee and Kasiri 2010) in a single system. This not only solves the often laborious photocatalysts recovery (Prado and Costa 2009) step, but also couples both hydrophobic (PHA) and hydrophilic (TiO_2) sites to attract mixed dye pollutants to be degraded photocatalytically.

6.4.3 Electrospinning of P(3HB)-TiO$_2$ Nanocomposite Fibers

The electrospinning conditions were optimized again after incorporating TiO_2 particles into the P(3HB) solution. Applied voltage and extrusion rate were set at 15 kV and 40 μL min^{-1}, respectively, for the electrospinning of P(3HB)-50 wt % TiO_2 precursor solution. Figure 6.5 shows the surface morphology of P(3HB)-TiO_2 fibers with different TiO_2 loading. The roughening of the fiber surface increased with catalyst loading as more TiO_2 particles were deposited on the polymer matrix. The electrospun fibers retained their homogeneous and cylindrical shape similar to neat P(3HB) fibers. This observation suggests that the addition of TiO_2 particles did not negatively alter the electrospinnability of 4 % (w/v) P(3HB) polymer solution. The average fiber diameter increased with TiO_2 loading from 10 to 40 wt% but a sharp decrease was observed at 50 wt% (Fig. 6.6). At 50 wt% TiO_2 loading, the entire nanofiber strands were saturated with TiO_2 particles. The average fiber diameter was 780 nm. When the TiO_2 concentration was increased to 60 wt%, electrospinning process was suppressed due to the high viscosity of the precursor solution.

As compared to the brittle cast P(3HB)-50 wt% TiO_2 film, the electrospun film of similar composition had a more flexible nature. This allows the electrospun P(3HB)-50 wt% TiO_2 to adapt various geometry. It has been reported that catalysts

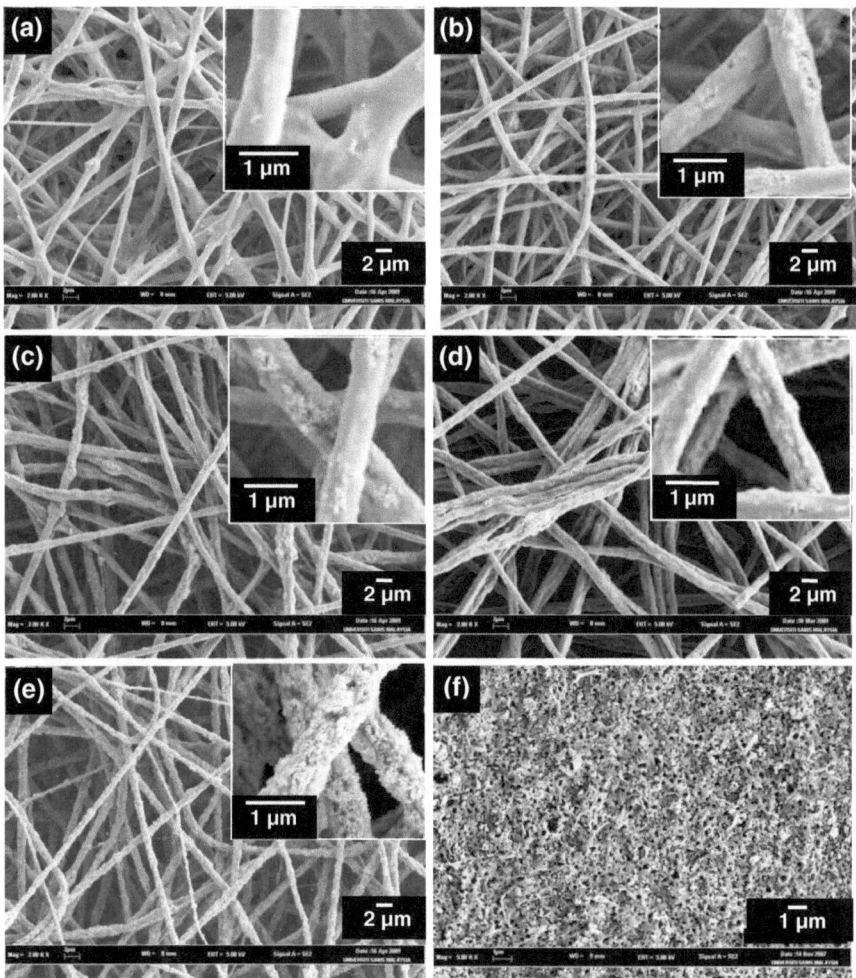

Fig. 6.5 SEM micrographs of fibers electrospun from P(3HB)- TiO$_2$ solution with **a** 10 % (w/w); **b** 20 % (w/w); **c** 30 % (w/w); **d** 40 % (w/w), and **e** 50 % (w/w) TiO$_2$ loading in 4 % (w/v) of P(3HB). The CHCl$_3$/DMF ratio used was 8:2. Applied voltage and extrusion rate were fixed at 15 kV and 40 μL min^{-1}, respectively. The morphology of cast P(3HB)-50 wt% TiO$_2$ is shown in **f**. Inset: magnified views

supported on nanofibrous substrates have better mechanical properties than fibrous catalysts. The intrinsic catalyst effect was reported to be more pronounced when loaded in ultrafine fibers (Subbiah et al. 2005). According to Zhang et al. (2009b), the electrospun nanofibers have high surface area which is approximately 1–2 orders of magnitude more than thin films. The cross-sectional view of the P(3HB)-50 wt % TiO$_2$ is shown in Fig. 6.7. P(3HB) fibers containing TiO$_2$ were found to be randomly deposited layer by layer forming a 3D open structure. The thickness

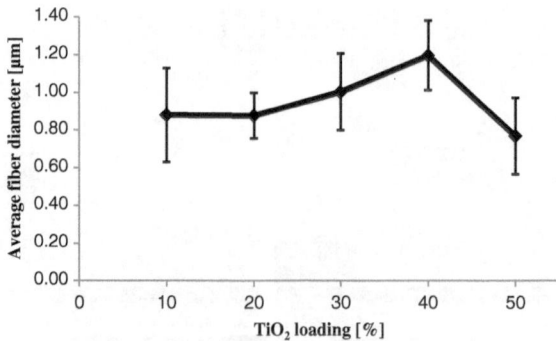

Fig. 6.6 Variation in average diameter of P(3HB)- TiO$_2$ fibers electrospun with varied TiO$_2$ loading. The ratio of CHCl$_3$/DMF mixed solvent used was 8:2. The applied voltage and extrusion rate were fixed at 15 kV and 40 μL min^{-1}, respectively

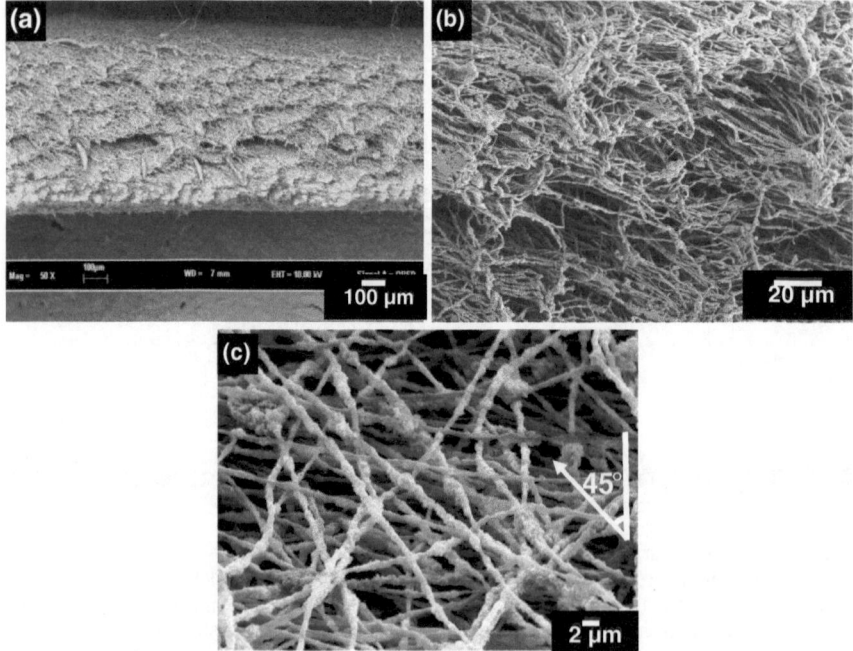

Fig. 6.7 SEM micrographs showing cross-sectional view of a five layered P(3HB)-50 wt% TiO$_2$ nanofibrous film electrospun at applied voltage and extrusion rate of 15 kV and 40 μL min^{-1}, respectively **a** 50 × and **b** 500 × magnifications. The ratio of CHCl$_3$/DMF used was 8:2. The morphology of the fibrous network from a tilted angle of 45° is shown in **c**

of the electrospun film was estimated to be approximately 0.51 mm. Pores of various diameters were formed by the differently oriented fibers stacked upon each other. It could be postulated that the fibers were completely dry upon reaching the collector as there were no coalesced junctions observed. This factor might have contributed to the larger thickness of the electrospun film. It is to be noted here that the cast film of similar composition has an approximate thickness of 0.25 mm, which is about half of the thickness of electrospun film.

6.4.4 Effect of CHCl₃/DMF Ratio on the Morphology of Electrospun P(3HB)-50 wt % TiO₂ Nanocomposite Fibers

In order to further improve the fiber morphology, the solvent concentrations were changed. The effects of $CHCl_3$/DMF ratio on the morphology of electrospun P(3HB)-50 wt% TiO_2 fibers were evaluated (Fig. 6.8). When neat $CHCl_3$ was used to prepare P(3HB)-50 wt % TiO_2 precursor solution, beaded fibers (beads on string) were obtained at 15 kV unlike incipient fibers that formed in the P(3HB)-$CHCl_3$ system at similar conditions. Fibers formed more readily in the [P(3HB)-TiO_2]-$CHCl_3$ system possibly due to the higher viscosity of the resultant precursor solution after TiO_2 addition. According to (Shenoy et al. 2005b), solid–liquid (S–L) phase separation (crystallization) can create additional intersection points, thereby lowering the concentration critical for fiber formation. Similar mechanism could have occurred in the [P(3HB)-TiO_2]-$CHCl_3$ system, thus enhancing polymeric chain entanglement due to the presence of TiO_2 particles. In addition, the solution conductivity might have been increased due to the semiconductive nature of TiO_2. However, the bead defects on the fibers were dominant. Agglomerates of TiO_2 particles were seen to be entrapped in the porous beads. This might be caused by the intrinsic agglomeration of TiO_2 and the effect of phase separation. Scarce amount of TiO_2 particles were immobilized on the fiber surface. This phenomenon is unfavorable as it will significantly reduce the available surface active sites of TiO_2. As expected, the incorporation of DMF as a co-solvent significantly eliminated bead defects. At $CHCl_3$/DMF ratios of 9:1 and 8:2, TiO_2 particles were dispersed very homogeneously on the fiber surface.

As the ratio of DMF was increased, the average fiber diameter reduced gradually (Fig. 6.9). This is in good agreement with previous report in which the polystyrene (PS) fiber diameter decreased with increasing DMF content in DMF/THF mixed solvent (Zheng et al. 2006). Although finer P(3HB)-TiO_2 fibers were formed at $CHCl_3$/DMF ratio of 7:3, TiO_2 particles were inhomogeneously distributed. The electrolytic DMF is known to randomly dissociate into positively and negatively charged ions in solution (Lee et al. 2003). Charge repulsion might have occurred between these ions and TiO_2 particles causing the latter to be scattered and aggregated. At a ratio of 6:4, the evaporation rate of the mixed solvent was greatly reduced due to the higher portion of DMF. The formed nanofibrous network was found to be coalesced at fiber junctions due to incomplete evaporation of solvent from the jet upon reaching the target. Almost no TiO_2 particle was observed on the fiber surface. Again, the more intense charge repulsion at 6:4 ratio might have caused the catalyst particles to be repelled from the surface of electrospun fibers. Thus 8:2 was set as the optimum ratio of $CHCl_3$/DMF mixed solvent to prepare the electrospun P(3HB)-50 wt % TiO_2.

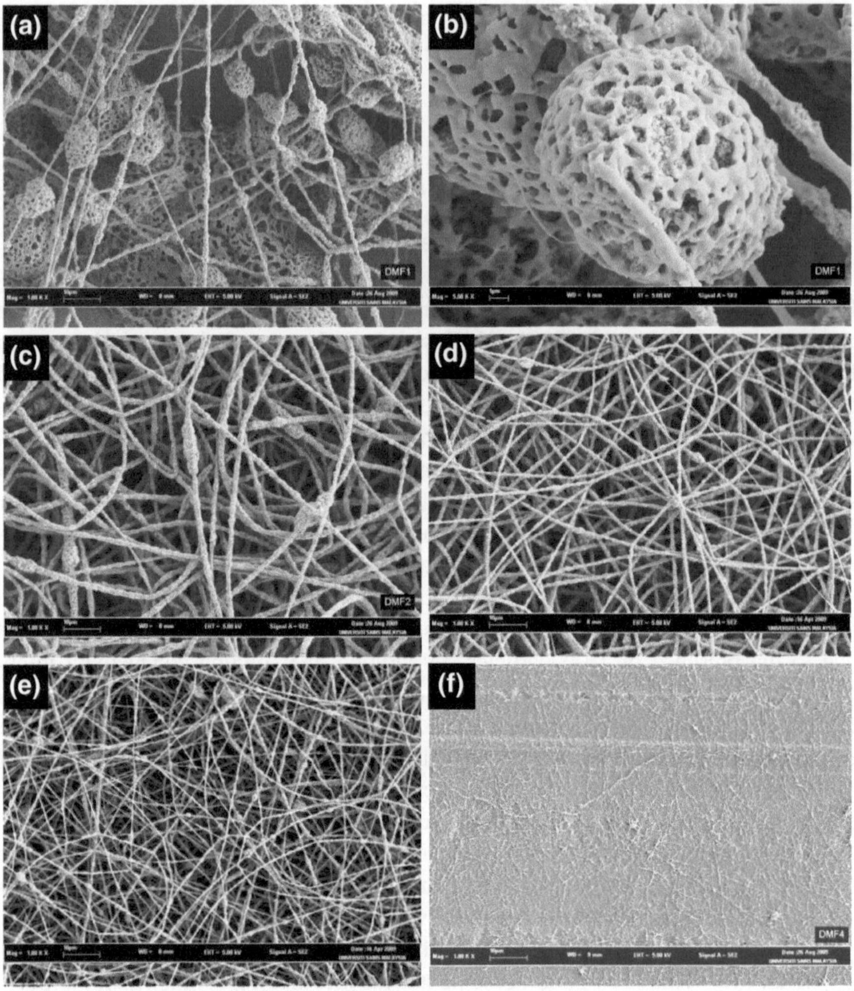

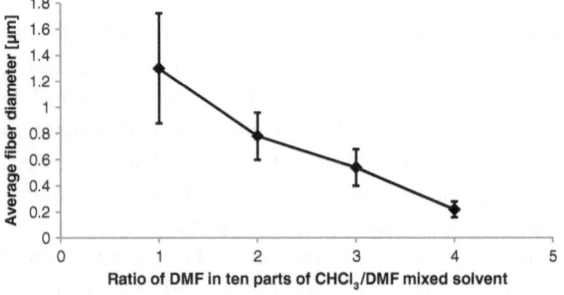

Fig. 6.8 Effect of CHCl₃/DMF mixed solvent ratio of **a** & **b** 10:0; **c** 9:1; **d** 8:2; **e** 7:3, and **f** 6:4 on the resultant P(3HB)-50 wt% TiO₂ electrospun fibers. The P(3HB) polymer concentration used was 4 % (w/v). Applied voltage and extrusion rate were fixed at 15 kV and 40 μL min⁻¹, respectively

Fig. 6.9 Variation in average diameter of P(3HB)-50 wt % TiO₂ fibers electrospun at different ratios of CHCl₃/DMF mixed solvent. The applied voltage and extrusion rate were fixed at 15 kV and 40 μL min⁻¹, respectively

Fig. 6.10 Morphological features of P(3HB)-50 wt % TiO$_2$ fibers electrospun from CHCl$_3$/DMF mixed solvent (ratio = 8:2) at **a** 15 kV; **b** 20 kV; **c** 25 kV; and **d** 30 kV with an extrusion rate of 40 µL min^{-1}. The P(3HB) polymer concentration used was 4 % (w/v)

6.4.5 Effect of Applied Voltage on the Morphology of Electrospun P(3HB)-50 wt % TiO$_2$ Nanocomposite Fibers

Further optimization was carried out by changing the applied voltage. The effect of applied voltage on the morphology of P(3HB)-50 wt% TiO$_2$ fibers is illustrated in Fig. 6.10. The TiO$_2$ distribution was found to be inhomogeneous in fibers electrospun at voltages above 15 kV. This might be due to the semiconductor particles being scattered as a result of higher charge repulsion. Most parts of the fibers were seen to be uncoated with TiO$_2$ particles as the nanoparticle agglomerates were centered at selective areas on the fibers. The average fiber diameter of P(3HB)-50 wt % TiO$_2$ was found to decrease as the applied voltage was increased from 15 to 30 kV (Fig. 6.11). This finding is in good agreement with previous studies where the average diameter of PLLA (Ren et al. 2008) and PS (Subbiah et al. 2005) nanofibers decreased with increasing voltage. When an electrically charged jet is ejected from the Taylor cone, whipping instability causes bending and stretching of the jet (Deitzel et al. 2001). When the radial charge repulsion overcomes cohesive forces within a jet, the primary jet splits into subjets (Frenot and Chronakis 2003).

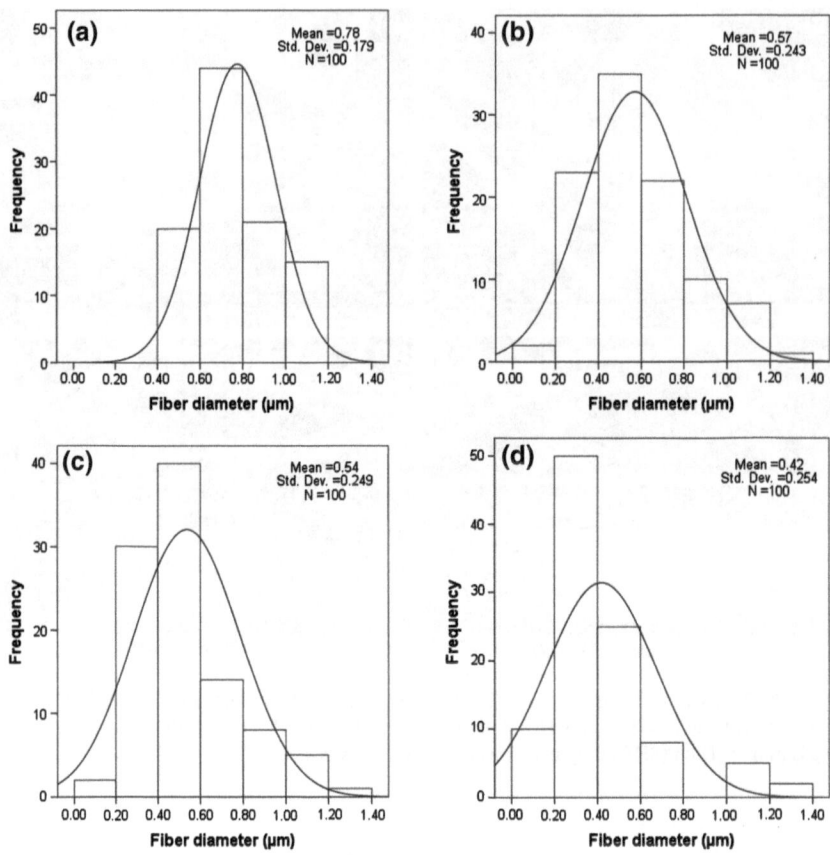

Fig. 6.11 Fiber diameter distribution of P(3HB)-50 wt % TiO$_2$ fibers electrospun from CHCl$_3$/DMF mixed solvent (ratio = 8:2) at **a** 15 kV; **b** 20 kV; **c** 25 kV; and **d** 30 kV with an extrusion rate of 40 μL min^{-1}

At higher applied voltage, bending and stretching of the subjets are intensified due to increasing repulsive forces between the jets (Ren et al. 2008; Wang et al. 2008). This results in the thinning of the final resulting jets, thus forming ultrafine fibers. A narrow distribution of nanofiber diameters was obtained at 15 kV, whereas fibers electrospun at voltages of 20–30 kV had a broader distribution pattern (Fig. 6.11). Similar distribution broadening was observed for P(3HB-co-3HHx) fibers electrospun at 18 kV using CHCl$_3$/DMF system. Broad diameter distribution is often generated by imbalance in the jet splay's viscous interaction and surface tension (Cheng et al. 2008) due to high charge density in the jets (Subbiah et al. 2005). Although the average fiber diameter of P(3HB)-50 wt% TiO$_2$ fibers was higher at 15 kV, there was more homogeneous distribution of TiO$_2$ particles and uniformity in the fiber morphology. The narrow distribution of the fiber diameter obtained at 15 kV ensured a better quality control in the resultant

Table 6.2 Energy dispersive X-ray (EDX) analysis of electrospun and cast P(3HB)-50 wt % TiO_2 composite films

Element	Weight % of elements	
	Electrospun P(3HB)-50 wt% TiO_2	Cast P(3HB)-50 wt% TiO_2
Titanium	34.51 ± 0.91	34.19 ± 1.13
Oxygen	41.35 ± 1.30	40.53 ± 1.15
Carbon	24.14 ± 1.78	23.87 ± 1.14

Note: Statistical analysis using one-way ANOVA showed that there were no significant differences (P 0.05) in the amount of every element in both films at the 95 % confidence level

electrospun film. P(3HB)-50 wt% TiO_2 fibers with an average fiber diameter of 780 nm were able to completely decolorize Batik dye wastewater in less than 3 h under solar irradiation with a chemical oxygen demand (COD) removal of 74 %.

6.4.6 Bactericidal Assessment on Cast and Electrospun P(3HB)-50 wt % TiO$_2$ Nanocomposite Fibers

Besides decolorizing dye photocatalytically, P(3HB)-TiO_2 nanofibers could also be used in antimicrobial treatment. Previously, P(3HB)-TiO_2 films were proven to inactivate *E. coli* upon irradiation by UVA in aqueous system (Yew et al. 2006a). Laboratory-scale experiments suggested that the microbial susceptibility rate could be enhanced 1–2 orders of magnitude using the electrospun P(3HB)-TiO_2 films. Photocatalysis by TiO_2 particles is due to the generation of reactive oxygen species (ROS) such as O^{2-}, ^{-}OH, and H_2O_2 upon activation by light (Dodd and Jha 2009). According to previous studies, peroxidation of polyunsaturated phospholipids by ROS leads to the disruption of cell membrane (Sunada et al. 2003; Alrousan et al. 2009). This causes detrimental effects to the normal respiratory activity and oxidative phosphorylation, resulting in cell death (Maness et al. 1999). The bactericidal efficiency of the electrospun and cast P(3HB)-50 wt% TiO_2 films were evaluated against *E. coli* JM109. Elemental distribution on both films was examined to ensure valid comparison between the films. EDX analysis confirmed that there was no significant difference in the composition of titanium, oxygen, and carbon on both films (Table 6.2).

The initial concentration of *E. coli* JM109 used in this study was 3.6×10^8 CFU ml^{-1}. Both films exhibited highest bactericidal activity when exposed to UVA light as compared to fluorescent light. At each condition, the antibacterial effect induced by electrospun films were enhanced 1–2 orders of magnitude more than the cast films (Fig. 6.12). The survival curve of *E. coli* JM109 showed an exponential decline upon illumination by UVA light in the presence of electrospun P(3HB)-50 wt% TiO_2. At 3 h, approximately 95 % of the bacterial cells were inactivated by the electrospun P(3HB)-50 wt% TiO_2 film. Complete inactivation

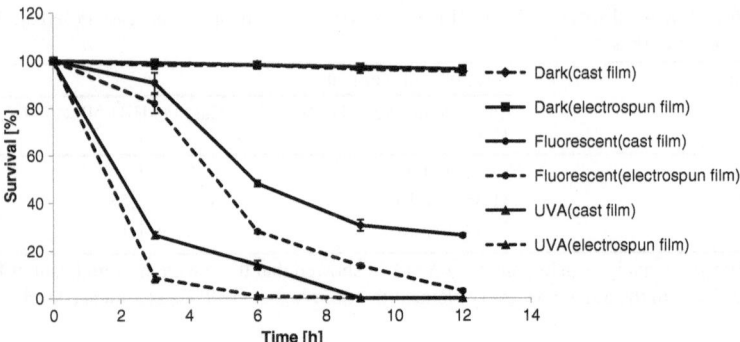

Fig. 6.12 Survival curve of *E. coli* JM 109 over exposure time to cast and electrospun P(3HB)-50 wt% TiO$_2$ films under different illumination sources. The initial cell concentration used was 3.6×10^8 CFU mL^{-1}. Bacterial suspension was sampled at 3 h intervals for 12 h. The samples were spread plated on LB agar and incubated at 37 °C for 24 h

of JM109 cells was achieved within 6 h. Whereas the cast P(3HB)-50 wt% TiO$_2$ films took almost 9 h to completely inactivate *E. coli* JM109 cells under UVA illumination. Under fluorescent light, electrospun P(3HB)-50 wt% TiO$_2$ film caused the survival of *E. coli* JM109 to decrease gradually and reached 100 % mortality after 12 h of illumination. The cast film on the other hand only caused 73 % mortality at the end of 12 h fluorescent illumination. The bactericidal activity of both electrospun and cast P(3HB)-50 wt % TiO$_2$ in the dark was negligible.

Photocatalytic activity depends on the contact of organic substrate with catalysts and better interfacial charge transfer reaction (Yang et al. 2010). Thus, the higher surface area and the 3D open structure of electrospun P(3HB)-50 wt% TiO$_2$ might have allowed more surface active sites of catalysts to be in contact with the bacterial cells. This could have yielded the enhanced bactericidal activity of electrospun P(3HB)-50 wt% TiO$_2$ films. TiO$_2$ particles are reported to be not toxic either in vitro or in vivo (Yew et al. 2006b). This knowledge coupled with the biocompatibility of P(3HB) (Tang et al. 2008) makes the electrospun P(3HB)-50 wt% TiO$_2$ film a potential disinfectant gauze to avoid wound infection. It could also serve as a self-sterillizing coating on medical devices. Ohko et al. (2001) reported coating of silicon catheter with TiO$_2$ via sol-gel method. It was proposed that the catheter could be sterilized by irradiation prior to insertion. Apart from this, the electrospun P(3HB)-50 wt% TiO$_2$ film could also be developed into an intelligent packaging material with antibacterial and self-cleaning properties to ensure the safety and quality of food products (Sozer and Kokini 2009). The biodegradability of P(3HB) (Manna and Paul 2000; Sridewi et al. 2006) is an added value for such packaging materials. It will also be an ecofriendly measure as most of the currently used packaging materials are non-degradable.

Chapter 7
Summary and Future Outlook

Abstract Polyhydroxyalkanoate (PHA) initially received serious attention as a possible substitute for petrochemical-based plastics because of the anticipated shortage in the supply of petroleum. Since then, PHA has remained as an interesting material to both the academia and industry. Now, we know more about this microbial storage polyester and have developed efficient fermentation systems for the large-scale production of PHA. Besides sugars, plant oils will become one of the important feedstock for the industrial-scale production of PHA. In addition, PHA will find new applications in various areas. This chapter summarizes the future prospects and the importance of developing a sustainable production system for PHA.

Keywords Palm oil • Plant oils • Polyhydroxyalkanoate (PHA) • Renewable resources • Sustainability

Advancement and continuous development in science and technology are undoubtedly essential for the well-being of mankind. We can never completely avoid the negative effects exerted onto our natural environment from the process of modernization. However, we can assure that the natural resources which fuel the modern discoveries and developments be consumed in a sustainable manner. This, in turn, will help to reduce damaging consequences on our precious ecosystems and natural environment. Sustenance is the key which ensures continuity and perseverance of natural resources and supplies for future generation. Synthetic plastics which are derived from finite resource have become an integral part of our lives. One can never foresee having to carry out daily chores without using at least one plastic-based product. Increasing demand for plastics in tandem with growing human population has resulted in the rapid accumulation of non-biodegradable materials in the environment.

A solution to this problem would be to reduce unnecessary usage of plastics and to recycle used plastic materials. These conservational efforts could be further aided by replacing some petrochemical-based plastics with biodegradable material possessing similar properties. One such potential candidate is PHA. This biodegradable polymer with thermoplastic properties is an ideal substitute for conventional plastics.

K. Sudesh, *Polyhydroxyalkanoates from Palm Oil: Biodegradable Plastics*, 101
SpringerBriefs in Microbiology, DOI: 10.1007/978-3-642-33539-6_7,
© The Author(s) 2013

Its properties are known to resemble some common plastics which are available commercially. The applications of PHA can also be diversified based on their properties. As described here in this book, usage of PHA in cosmetics and wastewater treatment highlights some niche applications for PHA. However, an important factor pertaining to the implementation of PHA globally is its high production cost. Carbon source used for bacterial fermentation of PHA has been identified as one of the major cost absorbing factors. Despite that, some PHA have been commercialized and marketed under different trademarks. However, these efforts have not been successful for long. Therefore, studies are still being devoted to identify cheap, renewable, and bio-based carbon feedstocks to further expand the global market of PHA products, thus, making it available to the public at a competitive price.

Unlike sugars which are currently used to produce most PHA, even at commercial scale, plant oils are being tested at laboratory level experiments for efficient PHA biosynthesis. Plant oils are known to generate higher PHA yields due to higher carbon content per gram of oil compared to sugars. To date, reports on the use of different plant oils including major commodity oils such as soybean and palm oil suggest that these oils could be considered as potential raw material for the production of various PHA at large scale. Among them, palm oil products are being studied extensively for the production of various types of PHA with improved and unique properties. It has been identified that high yield production of PHA could be obtained from palm oil and its by-products. Palm oil is a renewable and readily available resource in oil palm-rich countries such as Malaysia. Malaysia is the second largest producer of palm oil in the world after Indonesia. Various palm oil products and by-products have yielded positive results when used as carbon source for PHA biosynthesis. The studies provide preliminary results on the efficiency of palm oil bioconversion into PHA and future implementation of these substrates for larger and continuous PHA production systems.

Nevertheless, some important factors have to be considered if palm oil products are to be used as carbon feedstock for PHA production. It is expected that the demand for palm oil will continue to rise as a result of increasing human consumption. Hence, the expansion of oil palm plantation seems to be inevitable. However, proper land management and mitigation measures are necessary to avoid excessive deforestation and land wastage. Planters, industrialists, and researchers could work on further increasing the productivity of oil palms in already established plantations and improve the efficiency in palm oil recovery processes. This might reduce the need for new plantations. As for PHA production, usage of non-edible fractions of palm oils such as CPKO or the by-products such as PKAO and PAO as well as spent cooking oil will ensure uninterrupted supply of edible oil for palm oil-based food industries and human consumption. Utilization of waste materials from the milling process such as POME is also useful in converting waste to a value-added material. Development of an integrated system of PHA production with oil palm plantation and its milling industry could ensure a sustainable future for palm oil-based PHA industry. The future of PHA production from palm oil can be a success if the above measures are taken into consideration seriously in order to ensure a sustainable and environment-friendly process.

References

Abdelhad HM, Hafez AMA, El-sayed AA, Khodair TA (2009) Copolymer [P(HB-*co*-HV)] production as affected by strains and fermentation techniques. J Appl Sci Res 5:343–353

Abe C, Taima Y, Nakamura Y, Doi Y (1990) New bacterial copolyester of 3-hydroxyalkanoates and 3-hydroxy-ω-fluoroalkanoates produced by *Pseudomonas oleovorans*. Polym Commun 31:404–406

Abe H, Doi Y, Buschow KHJ, Robert WC, Merton CF, Bernard I (2001) Bacterial polyesters. Encyclopedia of materials: science and technology. Elsevier, Oxford. pp 448–453

Abe H, Kikkawa Y, Aoki H, Akehata T, Iwata T, Doi Y (1999) Crystallization behavior and thermal properties of melt-crystallized poly[(R)-3-hydroxybutyric acid-*co*-6-hydroxyhexanoic acid] films. Int J Biol Macromol 25:177–183

Abou-Zeid D-M, Müler R-J, Deckwer W-D (2001) Degradation of natural and synthetic polyesters under anaerobic conditions. J Biotechnol 86:113–126

Achten WMJ, Verchot L, Franken YJ, Mathijs E, Singh VP, Aerts R, Muys B (2008) Jatropha bio-diesel production and use. Biomass Bioenerg 32:1063–1084

Adam SEI, Magzoub M (1975) Toxicity of *Jatropha curcas* for goats. Toxicology 4:388–389

Adebowale KO, Adedire CO (2006) Chemical composition and insecticidal properties of the underutilized *Jatropha curcas* seed oil. Afr J Biotechnol 5:901–906

Aderibigbe AO, Johnson COLE, Makkar HPS, Becker K, Foidl N (1997) Chemical composition and effect of heat on organic matter- and nitrogen-degradability and some antinutritional components of Jatropha meal. Anim Feed Sci Technol 67:223–243

Agus J, Kahar P, Abe H, Doi Y, Tsuge T (2006) Molecular weight characterization of poly[(R)-3-hydroxybutyrate] synthesized by genectically engineered strains of *Escherichia coli*. Polym Degrad Stab 91:1138–1146

Ahmad AL, Ismail S, Bhatia S (2003) Water recycling from palm oil mill effluent (POME) using membrane technology. Desalination 157:87–95

Ahmed WA, Salimon J (2009) Phorbol ester as toxic constituents of tropical *Jatropha curcas* seed oil. Eur J Sci Res 31:429–436

Ahn WS, Park SJ, Lee SY (2001) Production of poly(3-hydroxybutyrate) from whey by cell recycle fed-batch culture of recombinant *Escherichia coli*. Biotechnol Lett 23:235–240

Akaraonye E, Keshavarz T, Roy I (2010) Production of polyhydroxyalkanoates: the future green materials of choice. J Chem Technol Biotechnol 85:732–743

Akintayo ET (2004) Characteristics and composition of *Parkia biglobbossa* and *Jatropha curcas* oils and cakes. Bioresour Technol 92:307–310

Akiyama M, Taima Y, Doi Y (1992) Production of poly(3-hydroxyalkanoates) by a bacterium of the genus *Alcaligenes* utilizing long-chain fatty acids. Appl Microbiol Biotechnol 37:698–701

Akiyama M, Tsuge T, Doi Y (2003) Environmental life cycle comparison of polyhydroxyal-kanoates produced from renewable carbon resources by bacterial fermentation. Polym Degrad Stab 80:183–194

Ali N, Suhaimi NS (2009) Performance evaluation of locally fabricated asymmetric nanofiltration membrane for Batik industry effluent. World Appl Sci J 5:46–52

Alias Z, Tan IKP (2005) Isolation of palm oil-utilising, polyhydroxyalkanoate (PHA)-producing bacteria by an enrichment technique. Bioresour Technol 96:1229–1234

Allen AD, Anderson WA, Ayorinde FO, Eribo BE (2010) Biosynthesis and characterization of copolymer poly(3HB-co-3HV) from saponified Jatropha curcas oil by Pseudomonas oleovorans. J Ind Microbiol Biotechnol 37:849–856

Allen AD, Anderson WA, Ayorinde F, Eribo BE (2011) Isolation and characterization of an extracellular thermoalkanophilic P(3HB-co-3HV) depolymerase from Streptomyces sp. IN1. Int Biodeterior Biodegrad 65:777–785

Alrousan DMA, Dunlop PSM, McMurray TA, Byrne JA (2009) Photocatalytic inactivation of E. coli in surface water using immobilised nanoparticle TiO₂ films. Water Res 43:47–54

Alvarez HM, Kalscheuer R, Steinbüchel A (1997) Accumulation of storage lipids in species of Rhodococcus and Nocardia and effect of inhibitors and polyethylene glycol. Fett/Lipid 99:239–246

Amara AAAF (2008) Polyhydroxyalkanoates: From basic research and molecular biology to application. IUM Eng J 9:37–73

Amirul AA, Yahya ARM, Sudesh K, Azizan MNM, Majid MIA (2008) Biosynthesis of poly(3-hydroxybutyrate-co-4-hydroxybutyrate) copolymer by Cupriavidus sp. USMAA1020 isolated from lake Kulim. Malays Bioresour Technol 99:4903–4909

Anderson AJ, Dawes EA (1990) Occurrence, metabolism, metabolic role, and industrial uses of bacterial polyhydroxyalkanoates. Microbiol Rev 54:450–472

Anderson AJ, Haywood GW, Dawes EA (1990) Biosynthesis and composition of bacterial poly(hydroxyalkanoates). Int J Biol Macromol 12:102–105

Annuar MSM, Tan IKP, Ibrahim S, Ramachandran KB (2007) Production of medium-chain-length Poly(3-hydroxyalkanoates) from crude fatty acids mixture by Pseudomonas putida. Food Bioprod Process 85:104–119

Aregheore EM, Becker K, Makkar HPS (2003) Detoxification of a toxic variety of Jatropha curcas using heat and chemical treatments, and preliminary nutritional evaluation with rats. S Pac J Nat Appl Sci 21:50–56

Arostegui SM, Aponte MA, Diaz E (1999) Bacterial polyesters produced by Pseudomonas oleovorans containing nitrophenyl groups. Macromolecules 32:2889–2895

Arpigny JL, Jaeger K-E (1999) Bacterial lipolytic enzymes: classification and properties. Biochem J 343:177–183

Asada Y, Miyake M, Miyake J, Kurane R, Tokiwa Y (1999) Photosynthetic accumulation of poly-(hydroxybutyrate) by cyanobacteria - the metabolism and potential for CO₂ recycling. Int J Biol Macromol 25:37–42

Ashby RD, Solaiman DKY (2008) Poly(hydroxyalkanoate) biosynthesis from crude alaskan pollock (theragra chalcogramma) oil. J Polym Environ 16:221–229

Atlas RM, Bartha R (1993) In microbial ecology: fundamentals and applications, 3rd edn. The Benjamin/Cummings Publishing Company, Redwood City, pp 39–43

Augustus GDPS, Jayabalan M, Seiler GJ (2002) Evaluation and bioinduction of energy components of Jatropha curcas. Biomass Bioenerg 23:161–164

Azad AM (2006) Fabrication of transparent alumina (Al₂O₃) nanofibers by electrospinning. Mater Sci Eng A 5:468–473

Badami AS, Kreke MR, Thompson MS, Riffle JS, Goldstein AS (2006) Effect of fiber diameter on spreading, proliferation, and differentiation of osteoblastic cells on electrospun poly(lactic acid) substrates. Biomaterials 27:596–600

Balaji R, Rekha N, Deecaraman M, Manikandan L (2009) Antimetastatic and antiproliferative activity of methanolic fraction of Jatropha curcas against B16F10 melanoma induced lung metastasis in C57BL/6 mice. Afr J Pharm Pharmacol 3:547–555

Barker M, Safford R, Burgner S, Edwards C (2009) Industrial uses for crops: bioplastics. Leaflet of HGCA project report no. 450: Industrial uses for crops: Markets for bioplastics

Barnard GN, Sanders JK (1989) The poly-β-hydroxybutyrate granule in vivo. A new insight based on NMR spectroscopy of whole cells. J Biol Chem 264:3286–3291

Basiron Y (2007) Palm oil production through sustainable plantations. Eur J Lipid Sci Technol 109:289–295

Basiron Y, Balu N, Chandramohan D (2004) Palm oil: the driving force of world oils and fats economy. Oil Palm Ind Econ J 4:1–7

Belay A, Kato T, Ota Y (1996) *Spirulina (Arthrospira)*: potential application as an animal feed supplement. J Appl Phycol 8:303–311

Bengtsson S, Pisco AR, Reis MAM, Lemos PC (2010) Production of polyhydroxyalkanoates from fermented sugar cane molasses by a mixed culture enriched in glycogen accumulating organisms. J Biotechnol 145:253–263

Berchmans HJ, Hirata S (2008) Biodiesel production from crude *Jatropha curcas* L. seed oil with a high content of free fatty acids. Bioresour Technol 99:1716–1721

Berck P, Bigman D (1993) The multiple dimensions of the world food problem. Food security and food inventories in developing countries. CAB International, Wallingford

Bhardwaj N, Kundu SC (2010) Electrospinning: a fascinating fiber fabrication technique. Biotechnol Adv 28:325–347

Bhubalan K (2010) Application of the polyhydroxyalkanoate synthase of *Chromobacterium* sp. USM2 for the production of biodegradable plastics. Ph.D dissertation. Universiti Sains Malaysia

Bhubalan K, Lee W-H, Loo C-Y, Yamamoto T, Tsuge T, Doi Y, Sudesh K (2008) Controlled biosynthesis and characterization of poly(3-hydroxybutyrate-*co*-3-hydroxyvalerate-*co*-3-hydroxyhexanoate) from mixtures of palm kernel oil and 3HV-precursors. Polym Degrad Stab 93:17–23

Bhubalan K, Rathi D-N, Abe H, Iwata T, Sudesh K (2010) Improved synthesis of P(3HB-co-3HV-co-3HHx) terpolymers by mutant *Cupriavidus necator* using the PHA synthase gene of *Chromobacterium* sp. USM2 with high affinity towards 3HV. Polym Degrad Stab 95:1436–1442

Bhubalan K, Yong KH, Kam YC, K. S (2010b) Cloning and expression of the PHA synthase gene from a locally isolated *Chromobacterium* sp. USM2. Malays J Microbiol 6:81–90

Binet R, Létoffé S, Ghigo JM, Delepelaire P, Wandersman C (1997) Protein secretion by Gram-negative bacterial ABC exporters - a review. Gene 192:7–11

Bluhm TL, Hamer GK, Marchessault RH, Fyfe CA, Veregin RP (1986) Isodimorphism in bacterial poly(β-hydroxybutyrate-*co*-β-hydroxyvalerate). Macromolecules 19:2871–2876

Boekema BKHL, Beselin A, Breuer M, Hauer B, Koster M, Rosenau F, Jaeger K-E, Tommassen J (2007) Hexadecane and tween 80 stimulate lipase production in *Burkholderia glumae* by different mechanisms. Appl Environ Microbiol 73:3838–3844

Bohmert K, Balbo I, Steinbuchel A, Tischendorf G, Willmitzer L (2002) Constitutive expression of the beta-ketothiolase gene in transgenic plants. A major obstacle for obtaining polyhydroxybutyrate-producing plants. Plant Physiol 128:1282–1290

Boland ED, Telemeco TA, Simpson DG, Wnek GE, Bowlin GL (2004) Utilizing acid pretreatment and electrospinning to improve biocompatibility of poly(glycolic acid) for tissue engineering. J Biomed Mater Res B 71:144–152

Borah B, Thakur PS, Nigam JN (2002) The influence of nutritional and environmental conditions on the accumulation of poly-β-hydroxybutyrate in *Bacillus mycoides* RLJ B-017. J Appl Microbiol 92:776–783

Braunegg G, Bona R, Koller M (2004) Sustainable polymer production. Polym-Plas Technol Eng 43:1779–1793

Braunegg G, Lefebvre G, Genser KF (1998) Polyhydroxyalkanoates, biopolyesters from renewable resources: physiological and engineering aspects. J Biotechnol 65:127–161

Braunegg G, Sonnleitner B, Lafferty RM (1978) A rapid gas chromatographic method for the determination of poly-β-hydroxybutyric acid in microbial biomass. Eur J Appl Microbiol Biotechnol 6:29–37

Brigham CJ, Budde CF, Mahan AE, Rha C, Sinskey AJ (2010) Elucidation of β-oxidation pathways in *Ralstonia eutropha* H16 by examination of global gene expression. J Bacteriol 192:5454–5464

Byrom D (1987) Polymer synthesis by microorganisms: technology and economics. Trends Biotechnol 5:246–250

Calabia BP, Tokiwa Y (2006) A novel PHB depolymerase from a thermophilic *Streptomyces* sp. Biotechnol Lett 28:383–388

Can-Aké R, Erosa-Rejón G, May-Pat F, Peña-Rodríguez LM, Peraza-Sánchez SR (2004) Bioactive terpenoids from roots and leaves of *Jatropha gaumeri*. Rev Soc Quím Méx 48:11–14

Carr NG (1966) The occurrence of poly-β-hydroxybutyrate in the blue-green alga, Chlorogloea fritschii. Biochim Biophys Acta 120:308–310

Carter C, Finley W, Fry J, Jackson D, Willis L (2007) Palm oil markets and future supply. Eur J Lipid Sci Technol 109:307–314

Castilho LR, Mitchell DA, Freire DMG (2009) Production of polyhydroxyalkanoates (PHAs) from waste materials and by-products by submerged and solid-state fermentation. Bioresour Technol 100:5996–6009

Cavalheiro JMBT, de Almeida MCMD, Grandfils C, da Fonseca MMR (2009) Poly (3-hydroxybutyrate) production by Cupriavidus necator using waste glycerol. Process Biochem 44:509–515

Chakravarty P, Mhaisalkar V, Chakrabarti T (2010) Study on poly-hydroxyalkanoate (PHA) production in pilot scale continuous mode wastewater treatment system. Bioresour Technol 101:2896–2899

Chan CK, Jalani BS, Araffin D (2003) Refining plantation technologies for sustainable production: the path to eco-economy. In: Proceedings of the PIPOC 2003 international palm oil Congress Malaysian Palm Oil Board, Kuala Lumpur, pp 283–303

Chanprateep S, Kulpreecha S (2006) Production and characterization of biodegradable terpolymer poly(3-hydroxybutyrate-*co*-3-hydroxyvalerate-*co*-4-hydroxybutyrate) by *Alcaligenes* sp. A-04. J Biosci Bioeng 101:51–56

Chanvel-Lesrat DJ, Pellen-Mussi P, Auroy P, Bonnaure-Mallet M (1999) Evaluation of the in vitro biocompatibility of various elastomers. Biomaterials 20:291–299

Chee J-Y, Tan Y, Samian M-R, Sudesh K (2010) Isolation and characterization of a *Burkholderia* sp. USM (JCM15050) capable of producing polyhydroxyalkanoate (PHA) from triglycerides, fatty acids and glycerols. J Polym Environ 18:584–592

Chen C, Dong L, Yu PHF (2006) Characterization and properties of biodegradable poly(hydroxyalkanoates) and 4,4-dihydroxydiphenylpropane blends: intermolecular hydrogen bonds, miscibility and crystallization. Eur Polym J. doi:10.1016/j.eurpol ymj.2006.1007.1005

Chen G-Q, Wu Q (2005) The application of polyhydroxyalkanoates as tissue engineering materials. Biomaterials 26:6565–6578

Chen GQ, Zhang G, Park SJ, Lee SY (2001) Industrial scale production of poly(3-hydroxybutyrate-co-3-hydroxyhexanoate). Appl Microbiol Biotechnol 57:50–55

Cheng M-L, Lin C–C, Su H-L, Chen P-Y, Sun Y-M (2008) Processing and characterization of electrospun poly(3-hydroxybutyrate-co-3-hydroxyhexanoate) nanofibrous membranes. Polymer 49:546–553

Choi J-C, Shin H-D, Lee Y-H (2003) Modulation of 3-hydroxyvalerate molar fraction in poly(3-hydroxybutyrate-*co*-3-hydroxyvalerate) using *Ralstonia eutropha* transformant co-amplifying *phbC* and NADPH generation-related *zwf* genes. Enzyme Microb Technol 32:178–185

Choi J, Lee SY (1999) Factors affecting the economics of polyhydroxyalkanoate production by bacterial fermentation. Appl Microbiol Biotechnol 51:13–21

Choi MH, Yoon SC (1994) Polyester biosynthesis characteristics of *Pseudomonas citronellolis* grown on various carbon sources, including 3-methyl-branched substrates. Appl Environ Microbiol 60:3245–3254

Choi MH, Yoon SC, Lenz RW (1999) Production of poly(3-hydroxybutyric acid-*co*-4-hydroxybutyric acid) and poly(4-hydroxybutyric acid) without subsequent degradation by Hydrogenophaga pseudoflava. Appl Environ Microbiol 65:1570–1577

Chowdhury AA (1963) Poly-β-hydroxybuttersaure abbauende bakterien und exo-enzyme. Arch Microbiol 47:167–200

Ciferri O, Tiboni O (1985) The biochemistry and industrial potential of *Spirulina*. Annu Rev Microbiol 89:503–526

Cima LG, Cima MJ (1996) Tissue regeneration matrices by solid free form fabrication techniques. US Patent 5518680

Cima LG, Vacanti JP, Vacanti C, Ingber D, Mooney D, Langer R (1991) Tissue engineering by cell transplantation using degradable polymer substrates. J Biomech Eng 113:143–151

Clarke E, Wiseman J (2005) Effects of variability in trypsin inhibitor content of soya bean meals on true and apparent ileal digestibility of amino acids and pancreas size in broiler chicks. Anim Feed Sci Technol 121:125–138

Corey JM, Gertz CC, Wang BS, Birrell LK, Johnson SL, Martin DC, Feldman EL (2008) The design of electrospun PLLA nanofiber scaffolds compatible with serum-free growth of primary motor and sensory neurons. Acta Biomaterials 4:863–875

Corley RHV, Gray BS (1976) Yield and yield components In: Corley RHV, Hardon JJ, Wood BJ (eds) Oil palm research. Elsevier, Amsterdam, pp 77–86

Cuesta C, Sánchez-Muniz FJ, Hernandez I (1991) Evaluation of nonpolar methyl esters by column and gas chromatography for the assessment of used frying oils. J Am Chem Soc 68:443–445

Curley JM, Hazer B, Lenz RW (1996) Production of poly(3-hydroxyalkanoates) containing aromatic substituents by *Pseudomonas oleovorans*. Macromolecules 29:1762–1766

Da Silva MG, Vargas H, Poley LH, Rodriguez RS, Baptista GB (2005) Structural impact of hydroxyvalerate in polyhydroxyalkanoates (PHA_{scl}) dense film monitored by XPS and photothermal methods. J Brazil Chem Soc 16:790–795

Davis JB (1964) Cellular lipids of a *Nocardia* grown on propane and n-butane. Appl. Microbiol Biotechnol 12:301–304

Dawes EA, Senior PJ (1973) The role and regulation of energy reserve polymers in microorganisms. Adv Microbiol Physiol 10:135–266

De Smet MJ, Eggink G, Witholt B, Kingma J, Wynberg H (1983) Characterization of intracellular inclusions formed by *Pseudomonas oleovorans* during growth on octane. J Bacteriol 154:870–878

Dee KC, Puleo DA, Bizios R (2002) Blood-biomaterial interaction and coagulation. In: Wiley J (ed) An introduction to tissue-biomaterial interactions. Wiley, New York, pp 37–87

Defoirdt T, Boon N, Sorgeloos P, Verstraete W, Bossier P (2009) Short-chain fatty acids and poly-β-hydroxyalkanoates: (new) biocontrol agents for a sustainable animal production. Biotechnol Adv 27:680–685

Degelau A, Scheper T, Bailey JE, Guske C (1995) Fluorometric measurement of poly-β-hydroxybutyrate in *Alcaligenes eutrophus* by flow cytometry and spectrofluorometry. Appl Microbiol Biotechnol 42:653–657

Deitzel JM, Kleinmeyer J, Harris D, Tan NCB (2001) The effect of processing variables on the morphology of electrospun nanofibers and textiles. Polymer 42:261–272

Demir MM, Yilgor I, Yilgor R, Yerman B (2002) Electrospinning of polyurethane fibers. Polymer 43:3303–3309

Deng Y, Zhao K, Zhang XF, Hu P, Chen GQ (2002) Study on the three-dimentional proliferation of rabbit articular cartilage-derived chondrocytes on polyhydroxyalkanoate scaffolds. Biomaterials 23:4049–4056

Devappa RK, Makkar HPS, Becker K (2010) Optimization of conditions for the extraction of phorbol esters from Jatropha oil. Biomass Bioenerg 34:1125–1133

Dharmsthiti S, Kuhasuntisuk B (1998) Lipase from *Pseudomonas aeruginosa* LP602: biochemical properties and application for wastewater treatment. J Ind Microbiol Biotechnol 21:75–80

Dobarganes MC, Márquez-Ruiz G (1998) Regulation of used frying fats and validity of quick tests for discarding the fats. Grasas Aceites 49:331–335

Dobarganes MC, Velasco J, Dieffenbacher A (2000) Determination of polar compounds, polymerized and oxidized triacylglycerols, and diacylglycerols in oils and fats (Technical Report). Pure Appl Chem 72:1563–1575

Dodd NJF, Jha AN (2009) Titanium dioxide induced cell damage: a proposed role of the carboxyl radical. Mutat Res 660:79–82

Doi Y (1990) Microbial polyesters. VCH, New York

Doi Y, Abe C (1990) Biosynthesis and characterization of a new bacterial copolyester of 3-hydroxyalkanoates and 3-hydroxy-w-chloroalkanoates. Macromol 23:3705–3707

Doi Y, Kitamura S, Abe H (1995) Microbial synthesis and characterization of poly(3-hydroxybutyrate-co-3-hydroxyhexanoate). Macromolecules 28:4822–4828

Doi Y, Kunioka M, Nakamura Y, Soga K (1986) ^{1}H and ^{13}C NMR analysis of poly(β-hydroxybutyrate) isolated from *Bacillus megaterium*. Macromolecules 19:1274–1276

Doi Y, Kunioka M, Nakamura Y, Soga K (1987a) Biosynthesis of copolyesters in *Alcaligenes eutrophus* H16 from ^{13}C-labeled acetate and propionate. Macromolecules 20:2988–2991

Doi Y, Kunioka M, Nakamura Y, Soga K (1988a) Nuclear magnetic resonance studies on unusual bacterial copolyesters of 3-hydroxybutyrate and 4-hydroxybutyrate. Macromolecules 21:2722–2727

Doi Y, Segawa A, Kunioka M (1989) Biodegradable poly(3-hydroxybutyrate-co-4-hydroxybutyrate) produced from γ-butyrolactone and butyric acid by Alcaligenes eutrophus. Polym Commun Guildf 30:169–171

Doi Y, Segawa A, Nakamura S, Kunioka MT (1990) Production of biodegradable copolyesters by *Alcaligenes eutrophus*. In: Dawes EA (ed) New biosynthetic biodegradable polymers of industrial interest from microorganisms. Kluwer Academic Publishers, Dordrecht, pp 37–48

Doi Y, Steinbüchel A (2001) Biopolymers: polyesters II vol 3B. Wiley-VCH, Weinheim

Doi Y, Tamaki A, Kunioka M, Soga K (1987b) Biosynthesis of terpolyesters of 3-hydroxybutyrate, 3-hydroxyvalerate and 5-hydroxyvalerate from 5-chloropentanoic and pentanoic acids. Makromol Chem Rapid Commun 8:631–635

Doi Y, Tamaki A, Kunioka M, Soga K (1988b) Production of copolyesters of 3-hydroxybutyrate and 3-hydroxyvalerate by *Alcaligenes eutrophus* from butyric and pentanoic acids. Appl Microbiol Biotechnol 28 (4):330–334 doi:10.1007/bf00268190

Donald PF (2004) Biodiversity impacts of some agricultural commodity production systems. Conserv Biol 18:17–38

Du GC, Chen J, Yu J, Lun S (2001a) Continuous production of poly-3-hydroxybutyrate by *Ralstonia eutropha* in a two-stage culture system. J Biotechnol 88:59–65

Du GC, Chen J, Yu J, Lun S (2001b) Feeding strategy of propionic acid for production of poly(3-hydroxybutyrate-co-3-hydroxyvalerate) with *Ralstonia eutropha*. Biochem Eng J 8:103–110

Duan B, Dong C, Yuan X, Yao K (2004) Electrospinning of chitosan solutions in acetic acid with poly(ethylene oxide). J Biomater Sci Polym Ed 15:797–811

Eda G, Shivkumar S (2007) Bead-to-fiber transition in electrospun polystyrene. J Appl Polym Sci 106:475–487

Eggink G, de Waard P, Huijberts GNM (1995) Formation of novel PHAs from long chain fatty acids. Can J Microbiol 41(13):14–21

Eggink G, van der Wal H, Huijberts G (1990) Production of poly-3-hydroxyalkanoates by *P. putida* during growth on long-chain fatty acids. In: Dawes EA (ed) Novel Biosynthetic Biodegradable Polymers of Industrial Interest from microorganisms. Kluwer Academic Publishers, Dordrecht, pp 37–48

EIBM (2006) National Energy Policies. (Online). (Accessed 7 Oct 2010). Available from: Energy Information Bureau Malaysia http://eib.org.my/index.php?page=article&item=99,124

Enomoto K, Ajioka M, Yamaguchi A (1994) Polyhydroxycarboxylic acid and preparation thereof. US Patent 5310865

Esimone CO, Nworu CS, Jackson CL (2008) Cutaneous wound healing activity of a herbal ointment containing the leaf extract of *Jatropha curcas L.* (Euphorbiaceae). Int J Appl Res Nat Prod 1:1–4

Ewering C, Lutke-Eversloh T, Luftmann H, Steinbuchel A (2002) Identification of novel sulfur-containing bacterial polyesters: Biosynthesis of poly(3-hydroxy-S-propyl-ω-thioalkanoates) containing thioether linkages in the side chains. Microbiol Mol Biol Rev 148:1397–1406

Fekkes P, Driessen AJM (1999) Protein targeting to the bacterial cytoplasmic membrane. Microbiol Mol Biol Rev 63:161–173

Feng L, Wang Y, Inagawa Y, Kasuya K, Saito T, Doi Y, Inoue Y (2004) Enzymatic degradation behavior of comonomer compositionally fractionated bacterial poly(3-hydroxybutyrate-co-3-hydroxyvalerate)s by poly(3-hydroxyalkanoate) depolymerases isolated from Ralstonia pickettii T1 and Acidovorax sp. TP4. Polym Degrad Stab 84:95–104

Feng L, Watanabe T, Wang Y, Kichise T, Fukuchi T, Chen G-Q, Doi Y, Inoue Y (2002) Studies on comonomer compositional distribution of bacterial poly(3-hydroxybutyrate-*co*-3-hydroxyhexanoate)s and thermal characteristics of their factions. Biomacromol 3:1071–1077

Fernandez D, Rodriguez E, Bassas M, Vinas M, Solanas AM, Llorens J, Marques AM, Manresa A (2005) Agro-industrial oily wastes as substrates for PHA production by the new strain *Pseudomonas aeruginosa* NCIB 40045: Effect of culture conditions. Biochem Eng J 26:159–167

Findlay RH, White DC (1983) Polymeric beta-hydroxyalkanoates from environmental samples and *Bacillus megaterium.* Appl Environ Microbiol 45:71–78

Fischer H, Erdmann S, Holler E (1989) An unusual polyanion from *Physarum polycephalum* that inhibits homologous DNA polymerase a in vitro. Biochemistry 28:5219–5226

Fitzherbert EB, Struebig MJ, Morel A, Danielsen F, Brühl CA, Donald PF, Phalan B (2008) How will oil palm expansion affect biodiversity? Trends Ecol Evol 23:538–545

Fletcher A (1993) PHA as natural, biodegradable polyesters. Plastics from bacteria and for bacteria. Springer, New York, pp 77–93

Flieger M, Kantorová M, Prell A, Řezanka T, Votruba J (2003) Biodegradable plastics from renewable sources. Folia Microbiol 48:27–44

Foidl N, Foidl G, Sanchez M, Mittelbach M, Hackel S (1996) *Jatropha curcas L.* as a source for the production of biofuel in Nicaragua. Bioresour Technol 58:77–82

Forson FK, Oduro EK, Hammond-Donkoh E (2004) Performance of jatropha oil blends in a diesel engine. Renew Energ 29:1135–1145

Freier T (2006) Biopolyesters in tissue engineering applications. Adv Polym Sci 203:1–61

Frenot A, Chronakis IS (2003) Polymer nanofibers assembled by electrospinning. Curr Opin Colloid Interface Sci 8:64–75

Fritzsche K, Lenz RW, Fuller R (1990a) An unusual bacterial polyester with a phenyl pendant group. Macromol Chem 191:1957–1965

Fritzsche K, Lenz RW, Fuller RC (1990b) Production of unsaturated polyesters by *Pseudomonas oleovorans.* Int J Biol Macromol 12:85–91

Fritzsche K, Lenz WR, Fuller RC (1990c) Bacterial polyesters containing branched poly(β-hydroxyalkanoate) units. Int J Biol Macromol 12:92–101

Fukui T, Doi Y (1997) Cloning and analysis of the poly(3-hydroxybutyrate-*co*-3-hydroxyhexanoate) biosynthesis genes of *Aeromonas caviae.* J Bacteriol 179:4821–4830

Fukui T, Doi Y (1998) Efficient production of polyhydroxyalkanoates from plant oils by Alcaligenes eutrophus and its recombinant strain. Appl Microbiol Biotechnol 49:333–336

Furukawa T, Matsusue Y, Yasunaga T, Shikinami Y, Okuno M, Nakamura T (2000) Biodegradation behavior of ultra-high-strength hydroxyapatite/poly(L-lactide) composite rods for internal fixation of bone fractures. Biomaterials 21:889–898

Ganapathy T, Murugesan K, Gakkhar RP (2009) Performance optimization of Jatropha biodiesel engine model using Taguchi approach. Appl Energ 86:2476–2486

Gandhi VM, Cherian KM, Mulky MJ (1995) Toxicological studies on ratanjyot oil. Food Chem Toxicol 33:39–42

Garcia B, Olivera ER, Minambres B, Fernandez-Valverde M, Canedo LM, Prieto MA, Garcia JL, Martinez M, Luengo JM (1999) Novel biodegradable aromatic plastics from a bacterial source. Genetic and biochemical studies on a route of the phenylacetyl-CoA catabolon. J Biol Chem 274:29228–29241

Geller BE (1996) Bacterial polyesters. Synthesis, properties, and application. Russ Chem Rev 65:725–734

Geng X, Kwon OH, Jang J (2005) Electrospinning of chitosan dissolved in concentrated acetic acid solution. Biomaterials 26:5427–5432

Gerngross TU, Snell KD, Peoples OP, Sinskey AJ (1994) Overexpression and purification of the soluble polyhydroxyalkanoate synthase from Alcaligenes eutrophus: evidence for a required posttranslational modification for catalytic activity. Biochem 33:9311–9320

Goel G, Makkar HPS, Francis G, Becker K (2007) Phorbol esters: structure, biological activity, and toxicity in animals. Int J Toxicol 26:279–288

Goh EM (1993) Specialty fats from palm and palm kernel oils. Selected readings on palm oil and its uses. PORIM

Goonasekera MM, Gunawardana VK, Jayasena K, Mohammed SG, Balasubramaniam S (1995) Pregnancy terminating effect of *Jatropha curcas* in rats. J Ethnopharmacol 47:117–123

Gorenflo V, Steinbüchel A, Marose S, Rieseberg M, Scheper T (1999) Quantification of bacterial polyhydroxyalkanoic acids by Nile red staining. Appl Microbiol Biotechnol 51:765–772

Gouda MK, Swellam AE, Omar SH (2001) Production of PHB by a *Bacillus megaterium* strain using sugarcane molasses and corn steep liquor as sole carbon and nitrogen sources. Microbiol Res 156:201–207

Griffith LG (2000) Emerging design principles on biomaterials and scaffolds for tissue engineering. Ann NY Acad Sci 961:83–95

Grochulski P, Li Y, Schrag JD, Bouthillier F, Smith P, Harrison D, Rubin B, Cygler M (1993) Insights into interfacial activation from an open structure of *Candida rugosa* lipase. J Biol Chem 268:12843–12847

Gross RA, Biological and chemical routes to modulate PHA structure. Lecture presented at the International Symposium on Bacterial PHA. Montreal, Canada

Gübitz GM, Mittelbach M, Trabi M (1999) Exploitation of the tropical oil seed plant *Jatropha curcas* L. Bioresour Technol 67:73–82

Guérin P, Renard E, Langlois V (2010) Degradation of natural and artificial poly[(R)-3-hydroxy-alkanoate]s: from biodegradation to hydrolysis. In: Chen G-Q (ed) Plastics from bacteria: natural function and applications, vol 14. Springer, Berlin

Gui MM, Lee KT, Bhatia S (2008) Feasibility of edible oil vs. non-edible oil vs. waste edible oil as biodiesel feedstock. Energy 33:1646–1653

Gunstone FD (ed) (2002) Vegetable oils in food technology: composition, properties and uses 1st edn. Blackwell Publishing Ltd, Oxford

Gupta N, Rathi P, Singh R, Goswami VK, Gupta R (2005a) Single-step purification of lipase from *Burkholderia multivorans* using polypropylene matrix. Appl Microbiol Biotechnol 67:648–653

Gupta P, Elkins C, Long TE, Wilkes GL (2005b) Electrospinning of linear homopolymers of poly(methyl methacrylate): exploring relationships between fiber formation, viscosity, molecular weight and concentration in a good solvent. Polymer 46:4799–4810

Gupta VK, Suhas (2009) Application of low-cost adsorbents for dye removal—A review. J Environ Manag 90:2313–2342

Haba E, Vidal-Mas J, Bassas M, Espuny MJ, Llorens J, Manresa A (2007) Poly 3-(hydroxyal-kanoates) produced from oily substrates by *Pseudomonas aeruginosa* 47T2 (NCBIM 40044): Effect of nutrients and incubation temperature on polymer composition. Biochem Eng J 35:99–106

Hahn SK, Chang YK, Lee SY (1995) Recovery and characterization of poly(3-hydroxybutyric acid) synthesized in *Alcaligenes eutrophus* and recombinant *Escherichia coli*. Appl Environ Microbiol 61:34–39

Han J, Qiu YZ, Liu DC, Chen GC (2004) Engineered *Aeromonas hydrophila* for enhanced production of poly(3-hydroxybutyrate-*co*-3-hydroxyhexanoate) with alterable monomers composition. FEMS Microbiol Lett 239:195–201

Hangii UJ (1990) Pilot scale production of PHB with *Alcaligenes latus*.. In: Dawes EA (ed) Novel biodegradable microbial polymers. Kluwer Academic, The Netherlands, pp 65–70

Harland BF, Morris ER (1995) Phytate: a good or a bad food component? Nutr Res 15:733–754

Hartley CWS (1989) The oil palm, 3rd edn. Longman, Harlow

Hassan MA, Karim MIA, Shirai Y, Inagaki M, Nakanishi K, Hashimoto K (1997a) Economic analysis on production of bacterial polyhydroxyalkanoates from palm oil mill effluent. J Chem Eng Jpn 30:751–755

Hassan MA, Shirai Y, Kusubayashi N, Karim MIA, Nakanishi K, Hashimoto K (1996) Effect of organic acid profiles during anaerobic treatment of palm oil mill effluent on the production of polyhydroxyalkanoates by *Rhodobacter sphaeroides*. J Ferment Bioeng 82:151–156

Hassan MA, Shirai Y, Kusubayashi N, Karim MIA, Nakanishi K, Hashimoto K (1997b) The production of polyhydroxyalkanoate from anaerobically treated palm oil mill effluent by *Rhodobacter sphaeroides*. J Ferment Bioeng 83:485–488

Hassan MA, Yacob S, Shirai Y, Hung Y-T (2006) Treatment of palm oil wastewaters. Waste treatment in the food processing industry. Taylor and Francis Group, LLC., Boca Raton, pp 101–117

Hazer B, Lenz RW, Fuller RC (1993) Biosynthesis of some new polyesters from methyl-branched alkanoic acids by *P. oleovorans*. Poster presented at the Second National Meeting of the Bio/Environmentally Degradable Polymer Society. Chicago

He W, Tian W, Zhang G, Chen GQ, Zhang Z (1998) Production of novel polyhydroxyalkanoates by *Pseudomonas stutzeri* 1317 from glucose and soybean oil. FEMS Microbiol Lett 169:45–49

Heller J (1996) Physic nut *Jatropha curcas* L. Promoting the conservation and use of underutilized and neglected crops, vol 1. Names of the species and taxonomy. Int Plant Genet Resour Inst, Rome

Hermawan S, Jendrossek D (2007) Microscopical investigation of poly(3-hydroxybutyrate) granule formation in *Azotobacter vinelandii*. FEMS Microbiol Lett 266:60–64

Hesselmann RPX, Fleischmann T, Hany R, Zehnder AJB (1999) Determination of polyhydroxyalkanoates in activated sludge by ion chromatographic and enzymatic methods. J Microbiol 35:111–119

Heydarkhan-Hagvall S, Schenke-Layland K, Dhanasopon AP, Rofail F, Smith H, Wu BM, Shemin R, Beygui RE, MacLellan WR (2008) Three-dimensional electrospun ECM-based hybrid scaffolds for cardiovascular tissue engineering. Biomaterials 29:2907–2914

Hocking PJ, Marchessault RH (1994) Biopolyesters. In: Griffin GJL (ed) Chemistry and technology of biodegradable polymers. Chapman & Hall, London, pp 48–96

Hoffmann N, Rehm BHA (2004) Regulation of polyhydroxyalkanoate biosynthesis in *Pseudomonas putida* and *Pseudomonas aeruginosa*. FEMS Microbiol Lett 237:1–7

Holmes PA (1988) Biologically produced (R)-3-hydroxyalkanoate polymers and copolymers In: Bassett DC (ed) Developments in crystalline polymers-2. Elsevier Applied Science, New York, pp 1–65

Holmes, PA, Wright, LF, Collins, SH (1981) European patent application 052.459.

Hoover R, Hughes T, Chung HJ, Liu Q (2010) Composition, molecular structure, properties, and modification of pulse starches: A review. Food Res Int 43:399–413

Hrabak O (1992) Industrial production of poly-β-hydroxybutyrate. FEMS Microbiol Rev 103:251–256

Huang M-X, Hou P, Wei Q, Xu Y, Chen F (2008) A ribosome-inactivating protein (curcin 2) induced from *Jatropha curcas* can reduce viral and fungal infection in transgenic tobacco. Plant Growth Regul 54:115–123

Huang Z-M, Zhang Y-Z, Kotaki M, Ramakrishna S (2003) A review on polymer nanofibers by electrospinning and their applications in nanocomposites. Compos Sci Technol 63:2223–2253

Huang ZM, Zhang YZ, Ramakrishna S, Lim CT (2004) Electrospinning and mechanical characterization of gelatin nanofibers. Polymer 45:5361–5368

Igbinosa OO, Igbinosa EO, Aiyegoro OA (2009) Antimicrobial activity and phytochemical screening of stem bark extracts from *Jatropha curcas* (Linn). Afr J Pharm Pharmacol 3:058–062

Iram SH, Cronan JE (2006) The $\hat{I}^2$-oxidation systems of Escherichia coli and Salmonella enterica are not functionally equivalent. J Bacteriol 188:599–608

Ishihara Y, Shimizu H, Shioya S (1996) Mole fraction control of poly(3-hydroxybutyric-*co*-3-hydroxyvaleric) acid in fed-batch culture of *Alcaligenes eutrophus*. J Ferment Bioeng. 81:422–428

Ishii D, Lee WK, Kasuya KI, Iwata T (2007) Fine structure and enzymatic degradation of poly[(R)-3-hydroxybutyrate] and stereocomplexed poly(lactide) nanofibers. J Biotechnol 132:318–324

Iwata T, Tsunoda K, Aoyagi Y, Kusaka S, Yonezawa N, Doi Y (2003) Mechanical properties of uniaxially cold-drawn films of poly([R]-3-hydroxybutyrate). Polym Degrad Stab 79:217–224

Jaafar ZM, Wong KH, Kamaruddin N (2003) Greener energy solutions for a sustainable future: issues and challenges for Malaysia. Energy Policy 31:1061–1072

Jacob GS, Garbow JR, Schaefer J (1986) Direct measurement of poly(β-hydroxybutyrate) in a pseudomonad by solid state ^{13}C NMR. J Biol Chem 261:16785–11678

Jaeger K-E, Dijkstra BW, Reetz MT (1999) Bacterial biocatalyst molecular biology, three dimensional structures and biotechnological applications of lipases. Annu Rev Microbiol 53:315–351

Jaeger K-E, Ransac S, Dijkstra BW, Colson C, Van Heuvel M, Misset O (1994) Bacterial lipases. FEMS Microbiol Rev 15:29–63

Jaeger K-E, Steinbüchel A, Jendrossek D (1995) Subtrate specificities of bacterial polyhydroxyalkanoate depolymerases and lipases: bacterial lipases hydrolyze poly(ω-hydroxyalkanoates). Appl Environ Microbiol 61:3113–3118

Jaeger KE, Eggert T (2002) Lipases for biotechnology. Curr Opin Biotechnol 13:390–397

Jain S, Sharma MP (2010) Prospects of biodiesel from Jatropha in India: a review. Renew Sustain Energy Rev 14:763–771

Jalili R, Hosseini SA, Morshed M (2005) The effects of operating parameters on the morphology of electrospun polyacrilonitrile. Iran Polym J 14:1074–1081

Jaqoc RB (1952) 'Deli' oil palm and early introductions of *Elaeis guineensis* to Malaysia. Malays Agric J 35:3–10

Jarute G, Kainz A, Schroll G, Baena JR, Lendl B (2004) On-line determination of the intracellular poly(β-hydroxybutyric acid) content in transformed *Escherichia coli* and glucose during PHB production using stopped-flow attenuated total reflection FT-IR spectrometry. Anal Chem 76:6353–6358

Jau MH, Yew SP, Toh PSY, Chong ASC, Chu WL, Phang SM, Najimudin N, Sudesh K (2005) Biosynthesis and mobilization of poly(3-hydroxybutyrate) [P(3HB)] by *Spirulina platensis*. Int J Biol Macromol 36:144–151

Jem KJ, van der Pol JF, de Vos S (2010) Microbial lactic acid, its polymer poly(lactic acid), and their industrial applications. In: Chen G-Q (ed) Plastics from bacteria: natural function and applications, vol 14., Microbiology MonographsSpringer, Berlin, pp 323–346

Jendrossek D (2005) Fluorescence microscopical investigation of poly(3-hydroxybutyrate) granule formation in bacteria. Biomacromolecules 6:598–603

Jendrossek D (2009) Polyhydroxyalkanoate granules are complex subcellular organelles (Carbonosomes). J Bacteriol 191:3195–3202

Jendrossek D, Handrick R (2002) Microbial degradation of polyhydroxyalkanoates. Annu Rev Microbiol 56:403–432

Jin HJ, Chen J, Karageorgiou V, Altman GH, Kaplan DL (2004) Human bone marrow stromal cell responses on electrospun silk fibroin mats. Biomaterials 25:1039–1047

Jin HJ, Fridrikh SV, Ruthledge GC, Kaplan DL (2002) Electrospinning Bombyx mori silk with poly(ethylene oxide). Biomacromolecules 3:1233–1239

Jing L, Fang Y, Ying X, Wenxing H, Meng X, Syed MN, Fang C (2005) Toxic impact of ingested Jatropherol-I on selected enzymatic activities and the ultrastructure of midgut cells in silkworm, *Bombyx mori* L. J Appl Entomol 129:98–104

Johnstone B (1990) A throw away answer. Far Eastern Econ Rev 147:62–63

Joseph B, Ramteke PW, Thomas G (2008) Cold active microbial lipases: some hot issues and recent developments. Biotechnol Adv 26:457–470

Jung K, Hany R, Rentsch D, Storni T, Egli T, Witholt B (2000) Characterization of new bacterial copolyesters containing 3-hydroxyoxoalkanoates and acetoxy-3-hydroxyalkanoates. Macromolecules 33:8571–8575

Juttner RR, Lafferty RM, Knackmuss HJ (1975) A simple method for the determination of poly-β-hydroxybutyric acid in microbial biomass. Eur J Appl Microbiol 1:233–237

Kadouri D, Jurkevitch E, Okon Y, Castro-Sowinski S (2005) Ecological and agricultural significance of bacterial polyhydroxyalkanoates. Crit Rev Microbiol 31:55–67

Kahar P, Tsuge T, Taguchi K, Doi Y (2004) High yield production of polyhydroxyalkanoates from soybean oil by Ralstonia eutropha and its recombinant strain. Polym Degrad Stab 83:79–86

Kang CK, Kusaka S, Doi Y (1995) Structure and properties of poly(3-hydroxybutyrate-co-4-hydroxybutyrate) produced by Alcaligenes latus. Biotechnol Lett 17:583–588

Kansal SK, Kaur N, Singh S (2009) Photocatalytic degradation of two commercial reactive dyes in aqueous phase using nanophotocatalysts. Nanoscale Res Lett 4:709–716

Karpushova A, Brümmer F, Barth S, Lange S, Schmid RD (2005) Cloning, recombinant expression and biochemical characterisation of novel esterases from Bacillus sp. associated with the marine sponge Aplysina aerophoba. Appl Microbiol Biotechnol 67:59–69

Karr DB, Waters JK, Emerich DW (1983) Analysis of poly-β-hydroxybutyrate in Rhizobium japonicum bacteroids by ion-exclusion high-pressure liquid chromatography and UV detection. Appl Environ Microbiol 46:1339–1344

Kasuya K-i, Inoue Y, Doi Y (1996) Adsorption kinetics of bacterial PHB depolymerase on the surface of polyhydroxyalkanoate films. Int J Biol Macrom 19:35–40

Kasuya K-i, Ohura T, Masuda K, Doi Y (1999) Substrate and binding specificities of bacterial polyhydroxybutyrate depolymerases. Int J Biol Macromol 24:329–336

Kato M, M. HJB, Kang CK, Fukui T, Doi Y (1996) Production of a novel copolyester of 3-hydroxybutyric acid and medium-chain-length 3-hydroxyalkanoic acids by Pseudomonas sp. 61-3 from sugars Appl Microbiol Biotechnol 45:363-370

Kaushik N, Kumar K, Kumar S, Kaushik N, Roy S (2007) Genetic variability and divergence studies in seed traits and oil content of Jatropha (Jatropha curcas L.) accessions. Biomass Bioenerg 31:497–502

Kawada J, Lütke-Eversloh T, Steinbuchel A, Marchessault RH (2003) Physical properties of microbial polythioesters: Characterization of poly(3-mercaptoalkanoates) synthesized by engineered Escherichia coli. Biomacromol 4:1698–1702

Kawata Y, Aiba SI (2010) Poly(3-hydroxybutyrate) production by isolated Halomonas sp. KM-1 using waste glycerol. Biosci Biotechnol Biochem 74:175–177

Keenan TM, Nakas JP, Tanenbaum SW (2006) Polyhydroxyalkanoate copolymers from forest biomass. J Ind Microbiol Biotechnol 33:616–626

Keenan TM, Tanenbaum SW, Stipanovic AJ, Nakas JP (2004) Production and characterization of poly-β-hydroxyalkanoate copolymers from Burkholderia cepacia utilizing xylose and levulinic acid. Biotechnol Prog 20:1697–1704

Kek Y-K, Lee W-H, Sudesh K (2008) Efficient bioconversion of palm acid oil and palm kernel acid oil to poly(3-hydroxybutyrate) by Cupriavidus necator. Can J Chem 86:533–539

Kek YK, Chang CW, Amirul AA, Sudesh K (2010) Heterologous expression of Cupriavidus sp. USMAA2-4 PHA synthase gene in PHB⁻4 mutant for the production of poly(3-hydroxybutyrate) and its copolymers. World J Microbiol Biotechnol 26:1595–1603 Kek YK, Sudesh K (Unpublished)

Kellerhals MB, Kessler B, Witholt B (1999a) Closed-loop control of bacterial high-cell-density fed-batch cultures: Production of mcl-PHAs by Pseudomonas putida KT2442 under single-substrate and cofeeding conditions. Biotechnol Bioeng 65:306–315

Kellerhals MB, Kessler B, Witholt B (1999b) Development of a closed-loop control system for production of medium-chain-length poly(3-hydroxyalkanoates) (mcl-PHAs) from bacteria. Macromol Symp 144:385–389

Kenmoku T, Sugawa E, Yano T, Imamura T (2002) Polyhydroxyalkanoate with (methylsulfanyl) phenoxy structure in side chain. EP 1275727B1

Kenmoku T, Sugawa E, Yano T, Nomoto T, Imamura T, Suzuki T, Honma T (2004) Polyhydroxyalkanoate and method of producing same, and &ohgr -(2-thienylsulfanyl) alkanoic acid and method of producing same. US Patent 6686439

Keshavarz T, Roy I (2010) Polyhydroxyalkanoates: bioplastics with a green agenda. Curr Opin Microbiol 13:321–326

Kessler B, Ren Q, De Roo G, Prieto MA, Witholt B (2001) Engineering of biological systems for the synthesis of tailor-made polyhydroxyalkanoates, a class of versatile polymers. Chimia 55:119–122

Khanna S, Srivastava AK (2005a) Recent advances in microbial polyhydroxyalkanoates. Process Biochem 40:607–619

Khanna S, Srivastava AK (2005b) Statistical media optimization studies for growth and PHB production by *Ralstonia eutropha*. Process Biochem 40:2173–2182

Khanna S, Srivastava AK (2007) Production of poly(3-hydroxybutyric-*co*-3-hydroxyvaleric aicd) having a high hydroxyvalerate content with valeric acid feeding. J Ind Microbiol Biotechnol 34:457–461

Khataee AR, Kasiri MB (2010) Photocatalytic degradation of organic dyes in the presence of nanostructured titanium dioxide: Influence of the chemical structure of dyes. J Mol Catal A: Chem 328:8–26

Khataee AR, Pons MN, Zahraa O (2009) Photocatalytic degradation of three azo dyes using immobilized TiO_2 nanoparticles on glass plates activated by UV light irradiation: Influence of dye molecular structure. J Hazard Mater 168:451–457

Kim DY, Kim Y, Rhee YH (1998) Bacterial poly(3-hydroxyalkanoates) bearing carbon–carbon triple bonds. Macromolecules 31:4760–4763

Kim DY, Kim YB, Rhee YH (2000) Evaluation of various carbon substrates for the biosynthesis of polyhydroxyalkanoates bearing functional groups by *Pseudomonas putida*. Int J Biol Macromol 28:23–29

Kim DY, Lutke-Eversloh T, Elbanna K, Thakor N, Steinbüchel A (2005) Poly(3-mercaptopropionate): A nonbiodegradable biopolymer? Biomacromolecules 6:897–901

Kim DY, Nam JS, Rhee YH, Kim YB (2003) Biosynthesis of novel poly(3-hydroxyalkanoates) containing alkoxy groups by *Pseudomonas oleovorans*. J Microbiol Biotechnol 13:632–635

Kim O, Gross RA, Hammar WJ, Newmark RA (1996a) Microbial synthesis of poly(β-hydroxyalkanoates) containing fluorinated side-chain substituents. Macromolecules 29:4572–4581

Kim YB, Lenz RW, Clinton Fuller R (1992) Poly(hydroxyalkanoate) copolymers containing brominated repeating units produced by Pseudomonas oleovorans. Macromolecules 25:1852–1857

Kim YB, Lenz RW, Fuller RC (1991) Preparation and characterization of poly(β-hydroxyalkanoates) obtained from Pseudomonas oleovorans grown with mixtures of 5-phenylvaleric acid and n-alkanoic acids. Macromolecules 24:5256–5260

Kim YB, Rhee YH, Heo GS, Kim JS (1996b) Poly-3-hydroxyalkanoates produced from *Pseudomonas oleovorans* grown with ω-Phenoxyalkanoates. Macromolecules 29:3432–3435

Kitamura S, Doi Y (1994) Staining method of poly(3-hydroxyalkanoic acids) producing bacteria by Nile blue. Biotechnol Tech 8:345–350

Kobayashi G, Tanaka K, Itoh H, Tsuge T, Sonomoto K, Ishizaki A (2000) Fermentative production of P(3HB-*co*-3HV) from propionic acid by *Alcaligenes eutrophus* in fed-batch culture with pH-stat continuous substrate feeding method. Biotechnol Lett 22:1067–1069

Koh LP (2008) Can oil palm plantations be made more hospitable for forest butterflies and birds? J Appl Ecol 45:1002–1009

Koh LP, Wilcove DS (2008) Is oil palm agriculture really destroying tropical biodiversity? Conserv Lett 1:60–64

Koller M, Atlic A, Gonzalez-Garcia Y, Kutschera C, Braunegg G (2008) Polyhydroxyalkanoate (PHA) biosynthesis from whey lactose. Macromol Symp 272:87–92

Koller M, Bona R, Braunegg G, Hermann C, Horvat P, Kroutil M, Martinz J, Neto J, Pereira L, Varila P (2005) Production of polyhydroxyalkanoates from agricultural waste and surplus materials. Biomacromol 6:561–565

Koller M, Hesse P, Bona R, Kutschera C, Atlic A, Braunegg G (2007) Potential of various archae- and eubacterial strains as industrial polyhydroxyalkanoate producers from whey. Macromol Biosci 7:218–226

Köse GT, Kenar H, Hasırcı N, Hasırcı V (2003) Macroporous poly(3-hydroxybutyrate-co-3-hydroxyvalerate) matrices for bone tissue engineering. Biomaterials 24:1949–1958

Kourmentza C, Ntaikou I, Kornaros M, Lyberatos G (2009) Production of PHAs from mixed and pure cultures of Pseudomonas sp. using short-chain fatty acids as carbon source under nitrogen limitation. Desalination 248:723–732

Kulkarni N, Gadre RV (1999) A novel alkaline, thermostable, protease-free lipase from Pseudomonas. Biotechnol Lett 21:897–899

Kulpreecha S, Boonruangthavorn A, Meksiriporn B, Thongchul N (2009) Inexpensive fed-batch cultivation for high poly(3-hydroxybutyrate) production by a new isolate of Bacillus megaterium. J Biosci Bioeng 107:240–245

Kumar A, Gross RA, Jendrossek D (2000) Poly(3-hydroxybutyrate)-depolymerase from Pseudomonas lemoignei: catalysis of esterifications in organic media. J Org Chem 65:7800–7806

Kumar A, Sharma S (2008) An evaluation of multipurpose oil seed crop for industrial uses (Jatropha curcas L.): a review. Ind Crop Prod 28:1–10

Kumar V, Rani A, Tindwani C, Jain M (2003) Lipoxygenase isozymes and trypsin inhibitor activities in soybean as influenced by growing location. Food Chem 83:79–83

Kunioka M, Kawaguchi Y, Doi Y (1989) Production of biodegradable copolyesters of 3-hydroxybutyrate and 4-hydroxybutyrate by Alcaligenes eutrophus. Appl Microbiol Biotechnol 30:56–573

Kunioka M, Nakamura Y, Doi Y (1988) New bacterial copolyesters produced in Alcaligenes eutrophus from organic acids. Polym commun 29:174–176

Kuo CY, Lin HY (2009) Photodegration of C.I. Reactive Red 2 by platinized titanium dioxide. J Hazard Mater 165:1243–1247

Kusaka S, Iwata T, Doi Y (1998) Microbial synthesis and physical properties of ultra-high-molecular-weight poly[(R)-3-hydroxybutyrate]. J Macromol Sci - Pure and Appl Chem 35:319–335

Kusaka S, Iwata T, Doi Y (1999) Properties and biodegradability of ultra-high-molecular-weight poly[(R)-3-hydroxybutyrate] produced by a recombinant Escherichia coli. Int J Biol Macromol 25:87–94

Łabużek S, Radecka I (2001) Biosynthesis of PHB tercopolymer by Bacillus cereus UW85. J Appl Microbiol 90:353–357

Lageveen RG, Huisman GW, Preusting H, Ketelaar P, Eggink G, Witholt B (1988) Formation of polyesters by Pseudomonas oleovorans: effect of substrates on formation and composition of poly-(R)-3-hydroxyalkanoates and poly-(R)-3-hydroxyalkenoates. Appl. Environ. Microbiol. 54:2924–2932

Lam MK, Tan KT, Lee KT, Mohamed AR (2009) Malaysian palm oil: Surviving the food versus fuel dispute for a sustainable future. Renew Sustain Energy Rev 13:1456–1464

Lan GQ, Abdullah N, Jalaludin S, Ho YW (2010) In vitro and in vivo enzymatic dephosphorylation of phytate in maize-soya bean meal diets for broiler chickens by phytase of Mitsuokella jalaludinii. Anim Feed Sci Technol 158:155–164

Lawrence AG, Schoenheit J, He A, Tian J, Liu P, Stubbe J, Sinskey AJ (2005) Transcriptional analysis of Ralstonia eutropha genes related to poly-(R)-hydroxybutyrate homeostasis during batch fermentation. Appl Microbiol Biotechnol 68:663–672

Lee C-H, Parkin KL (2003) FA selectivity of lipases in acyl-transfer reactions with acetate esters of polyols in organic media. J Am Oil Chem Soc 80(3):231–236

Lee EY, Jendrossek D, Schirmer A, Choi CY, Steinbuchel A (1995) Biosynthesis of copolyesters consisting of 3-hydroxybutyrate and medium-chain-length 3-hydroxyalkanoates from 1,3-butanediol or from 3-hydroxybutyrate by *Pseudomonas* sp. A33. Appl Microbiol Biotechnol 42:901–909

Lee IY, Kim GJ, Choi DK, Yeon BK, Park YH (1996) Improvement of hydroxyvalerate fraction in poly(β-hydroxybutyrate-*co*-β-hydroxyvalerate) by a mutant strain of *Alcaligenes eutrophus*. J Ferment Bioeng 81:255–258

Lee KH, Kim HY, Bang HJ, Jung YH, Lee SG (2003) The change of bead morphology formed on electrospun polystyrene fibers. Polymer 44:4029–4034

Lee SY (1996a) Bacterial polyhydroxyalkanoates. Biotechnol Bioeng 49:1–14

Lee SY (1996b) Plastic bacteria progress and prospects for polyhydroxyalkanoate production in bacteria. Trends Biotechnol 14:431–438

Lee SY, J-i Choi (1999) Production and degradation of polyhydroxyalkanoates in waste environment. Waste Manag 19:133–139

Lee SY, Choi JI (1998) Effect of fermentation performance on the economics of poly(3-hydroxy-butyrate) production by Alcaligenes latus. Polym Degrad Stab 59:387–393

Lee SY, Choi J, Wong HH (1999) Recent advances in polyhydroxyalkanoate production by bacterial fermentation: mini-review. Int J Biol Macromol 25:31–36

Lee SY, Lee KM, Chan HN, Steinbüchel A (1994) Comparison of recombinant *Escherichia coli* strains for synthesis and accumulation of poly(3-hydroxybutyric acid) and morphological changes. Biotechnol Bioeng 44:1337–1347

Lee W-H, Azizan MNM, Sudesh K (2004) Effects of culture conditions on the composition of poly(3-hydroxybutyrate-*co*-4-hydroxybutyrate) synthesized by *Comamonas acidovorans*. Polym Degrad Stab 84:129–134

Lee W-H, Loo C-Y, Nomura CT, Sudesh K (2008) Biosynthesis of polyhydroxyalkanoate copolymers from mixtures of plant oils and 3-hydroxyvalerate precursors. Bioresour Technol 99:6844–6851

Lemoigne M (1926) Produits de deshydration et de polymerisation de l'acide β-oxybutyric. Bull Soc Chim Biol 8:770–782

Lenz RW, Kim YB, Fuller RC (1992) Production of unusual bacterial polyesters by *Pseudomonas oleovorans* through cometabolism. FEMS Microbiol Lett 103:207–214

Lertsathapornsuk V, Pairintra R, Aryusuk K, Krisnangkura K (2008) Microwave assisted in continuous biodiesel production from waste frying palm oil and its performance in a 100 kW diesel generator. Fuel Process Technol 89:1330–1336

Li C-Y, Devappa RK, Liu J-X, Lv J-M, Makkar HPS, Becker K (2010) Toxicity of *Jatropha curcas* phorbol esters in mice. Food Chem Toxicol 48:620–625

Li D, Xia Y (2004) Electrospinning of nanofibers: reinventing the wheel? Adv Mater 16:1151–1170

Li R, Zhang H, Qi Q (2007) The production of polyhydroxyalkanoates in recombinant Escherichia coli. Bioresour Technol 98:2313–2320

Li WJ, Cooper JA, Mauck RL, Tuan RS (2006) Fabrication and characterization of six electrospun poly(alpha-hydroxy ester)-based fibrous scaffolds for tissue engineering applications. Acta Biomaterials 2:377–385

Li X-T, Zhang Y, Chen G-Q (2008) Nanofibrous polyhydroxyalkanoate matrices as cell growth supporting materials. Biomaterials 29:3720–3728

Lim S, Teong LK (2010) Recent trends, opportunities and challenges of biodiesel in Malaysia: an overview. Renew Sustain Energy Rev 14:938–954

Lim YY, Sudesh K (Unpublished).

Lin J, Yan F, Tang L, Chen F (2003) Antitumor effects of curcin from seeds of *Jatropha curcas*. Acta Pharmacol Sinica 24:241–246

Liu JH, Jen HL, Chung YC (1999) Surface modification of polyethylene membrans using phosphorylcholine derivatives and their platelet compatibility. J Appl Polym Sci 74:2947–2954

Loh SK, Choo YM, Cheng SF, Ma A (2006) Recovery and conversion of palm olein-derived used frying oil to methyl esters for biodiesel. J Oil Palm Res 18:247–252

Loo C-Y, Lee W-H, Tsuge T, Doi Y, Sudesh K (2005) Biosynthesis and characterization of poly(3-hydroxybutyrate-*co*-3-hydroxyhexanoate) from palm oil products in a *Wautersia eutropha* mutant. Biotechnol Lett 27:1405–1410

Loo C-Y, Sudesh K (2007) Biosynthesis and native granule characteristics of poly(3-hydroxybutyrate-*co*-3-hydroxyvalerate) in *Delftia acidovorans*. Int J Biol Macromol 40:466–477

Lu H, Liu Y, Zhou H, Yang Y, Chen M, Liang B (2009) Production of biodiesel from *Jatropha curcas* L. oil. Comput Chem Eng 33:1091–1096

Luengo JM, Garcíía B, Sandoval A, Naharro G, Olivera ER (2003) Bioplastics from microorganisms. Curr Opin Microbiol 6:251–260

Luo S, Grubb DT, Netravali AN (2002) The effect of molecular weight on the lamellar structure, thermal and mechanical properties of poly(hydroxybutyrate-*co*-hydroxyvalerates). Polymer 43:4159–4166

Lutke-Eversloh T, Bergander K, Luftmann H, Steinbuchel A (2001) Biosynthesis of poly (3-hydroxybutyrate-co-3-mercaptobutyrate) as a sulfur analogue to poly(3-hydroxybutyrate) (PHB). Biomacromolecules 2:1061–1065

Lutke-Eversloh T, Fischer A, Remminghorst U, Kawada J, Marchessault RH, Bogershausen A, Kalwei M, Eckert H, Reichelt R, Liu SJ, Steinbuchel A (2002) Biosynthesis of novel thermoplastic polythioesters by engineered *Escherichia coli*. Nat Mater 1:236–239

Lutke-Eversloh T, Steinbuchel A (1999) Biochemical and molecular characterization of a succinate semialdehyde dehydrogenase involved in the catabolism of 4-hydroxybutyric acid in Ralstonia eutropha. FEMS Microbiol Lett 181:63–71

Luzier WD (1992) Materials derived from biomass/biodegradable materials. Proc Natl Acad Sci USA 89:839–842

Madden LA, Anderson AJ, Asrar J, Berger P, Garrett P (2000) Production and characterization of poly(3-hydroxybutyrate-*co*-3-hydroxyvalerate-*co*-4-hydroxybutyrate) synthesized by *Ralstonia eutropha* in fed-batch cultures. Polymer 41:3499–3505

Madden LA, Anderson AJ, Shah DT, Asrar J (1999) Chain termination in polyhydroxyalkanoate synthesis: involvement of exogenous hydroxy-compounds as chain transfer agents. Int J Biol Macromol 25:43–53

Madison L, Huisman G (1999) Metabolic engineering of poly(3-hydroxyalkanoates): From DNA to plastic. Microbiol Mol Biol Rev 63:21–53

Mahanta N, Gupta A, Khare SK (2008) Production of protease and lipase by solvent tolerant *Pseudomonas aeruginosa* PseA in solid-state fermentation using *Jatropha curcas* seed cake as substrate. Bioresour Technol 99:1729–1735

Majid MIA, Akmal DH, Few LL, Agustien A, Toh MS, Samian MR, Najimudin N, Azizan MN (1999) Production of poly(3-hydroxybutyrate) and its copolymer poly(3- hydroxybutyrate-co-3-hydroxyvalerate) by Erwinia sp. USMI-20. Int J Biol Macromol 25:95–104

Majid MIA, Hori K, Akiyama M, Doi Y (1994) Production of poly(3-hydroxybutyrate) from plant oils by *Alcaligenes* sp. Stud Polym Sci 12:417–424

Makkar HPS, Aderibigbe AO, Becker K (1998) Comparative evaluation of non-toxic and toxic varieties of *Jatropha curcas* for chemical composition, digestibility, protein degradability and toxic factors. Food Chem 62:207–215

Makkar HPS, Becker K (1997) Potential of *J. curcas* seed meal as a protein supplement to livestock feed, constraints to its utilisation and possible strategies to overcome constraints. In: Gübits GM, Mittelbach M, Trabi M (eds) Biofuels and industrial products from *J. curcas*Dbv, Graz, pp 190–205

Makkar HPS, Becker K (1999) Nutritional studies on rats and fish (carp *Cyprinus carpio*) fed diets containing unheated and heated *Jatropha curcas* meal of a non-toxic provenance. Plant Food Hum Nutr 53:183–192

Makkar HPS, Becker K (2009) *Jatropha curcas*, a promising crop for the generation of biodiesel and value-added coproducts. Eur J Lipid Sci Technol 111:773–787

Makkar HPS, Martínez-Herrera J, Becker K (2008) Variations in seed number per fruit, seed physical parameters and contents of oil, protein and phorbol ester in toxic and non-toxic genotypes of *Jatropha curcas*. J Plant Sci 3:260–265

Maness PC, Smolinski S, Blake DM, Huang Z, Wolfrum EJ, Jacoby WA (1999) Bactericidal activity of photocatalytic TiO_2 reaction: toward an understanding of its killing mechanism. Appl Environ Microbiol 65:4094–4098

Manna A, Paul AK (2000) Degradation of microbial polyester poly(3-hydroxybutyrate) in environmental samples and in culture. Biodegradation 11:323–329

Marangoni C, Furigo A, Falcão de Aragão GM (2000) Oleic acid improves poly(3-hydroxybutyrate-*co*-3-hydroxyvalerate) production by *Ralstonia eutropha* in inverted sugar and propionic acid. Biotechnol Lett 22:1635–1638

Marsudi S, Unno H, Hori K (2008) Palm oil utilization for the simultaneous production of polyhydroxyalkanoates and rhamnolipids by *Pseudomonas aeruginosa*. Appl Microbiol Biotechnol 78:955–961

Martínez-Herrera J, Martínez-Ayala AL, Makkar H, Francis G, Becker K (2010) Agroclimatic conditions, chemical and nutritional characterization of different provenances of *Jatropha curcas* L. from Mexico. Eur J Sci Res 39:396–407

Martínez-Herrera J, Siddhuraju P, Francis G, Dávila-Ortíz G, Becker K (2006) Chemical composition, toxic/antimetabolic constituents, and effects of different treatments on their levels, in four provenances of *Jatropha curcas* L. from Mexico. Food Chem 96:80–89

Matavulj M, Molitoris HP (1992) Fungal degradation of polyhydroxyalkanoates and a semi-quantitative assay for screening their degradation by terrestrial fungi. FEMS Microbiol Lett 103:323–331

Matsumoto K, Arai Y, Nagao R, Murata T, Takase K, Nakashita H, Taguchi S, Shimada H, Doi Y (2006) Synthesis of short-chain-length/Medium-chain-length polyhydroxyalkanoate (PHA) copolymers in peroxisome of the transgenic Arabidopsis Thaliana harboring the PHA synthase gene from Pseudomonas sp. 61-3. J Polym Environ 14:369–374

Matsumoto K, Nagao R, Murata T, Arai Y, Kichise T, Nakashita H, Taguchi S, Shimada H, Doi Y (2005) Enhancement of poly(3-hydroxybutyrate-co-3-hydroxyvalerate) production in the transgenic Arabidopsis thaliana by the in vitro evolved highly active mutants of polyhydroxyalkanoate (PHA) synthase from Aeromonas caviae. Biomacromol 6:2126–2130

Matsusaki H, Abe H, Doi Y (2000) Biosynthesis and properties of poly(3-hydroxybutyrate-*co*-3-hydroxyalkanoates) by recombinant strains of *Pseudomonas* sp. 61-3. Biomacromol 1:17–22

Matsuse IT, Lim YA, Hattori M, Correa M, Gupta MP (1998) A search for anti-viral properties in Panamanian medicinal plants.: The effects on HIV and its essential enzymes. J Ethnopharmacol 64:15–22

Mauclaire L, Brombacher E, Zinn M, Bünger JD (2010) Factors controlling bacterial attachment and biofilm formation on medium-chain-length polyhydroxyalkanoates (mcl-PHAs). Colloids Surf B: Biointerfaces 76:104–111

Maycock JM (1992) Palm oil milling technology. Malays Oil Sci Technol (MOST) 1:31–35

McCool GJ, Cannon MC (2001) PhaC and PhaR are required for polyhydroxyalkanoic acid synthase activity in *Bacillus megaterium*. J Bacteriol 183:4235–4243

Meng W, Xing Z-C, Jung K-H, Kim S-Y, Yuan J, Kang I-K, Yoon S, Shin H (2008) Synthesis of gelatin-containing PHBV nanofiber mats for biomedical application. J Mater Sci: Mater Med 19:2799–2807

Mergaert J, Webb A, Anderson C, Wouters A, Swings J (1993) Microbial degradation of poly(3-hydroxybutyrate) and poly(3-hydroxybutyrate-co-3-hydroxyvalerate) in soils. Appl Environ Microbiol 59:3233–3238

Mifune J, Nakamura S, Fukui T (2008) Targeted engineering of Cupriavidus necator chromosome for biosynthesis of poly(3-hydroxybutyrate-co-3-hydroxyhexanoate) from vegetable oil. Can J Chem 86:621–627

Min BM, Lee G, Kim SH, Nam YS, Lee TS, Park WH (2004a) Electrospinning of silk fibroin nanofibers and its effect on the adhesion and spreading of normal human kerotinocytes and fibroblasts in vitro. Biomaterials 25:1289–1297

Min BM, Lee SW, Lim JN, You Y, Lee TS, Kang PH, Park WH (2004b) Chitin and chitosan nanofibers: electrospinning of chitin and deacetylation of chitin nanofibers. Polymer 45:7137–7142

Mitomo H, Hsieh WC, Nishiwaki K, Kasuya K, Doi Y (2001) Poly(3-hydroxybutyrate-co-4-hydroxybutyrate) produced by Comamonas acidovorans. Polymer 42:3455–3461

Mitomo H, Morishita N, Doi Y (1993) Composition range of crystal phase transition of isodimorphism in poly(3-hydroxybutyrate-*co*-3-hydroxyvalerate). Macromolecules 26:5809–5811

Mitomo H, Morishita N, Doi Y (1995) Structural changes of poly(3-hydroxybutyrate-*co*-3-hydroxyvalerate) fractionated with acetone-water solution. Polymer 36:2573–2578

Mitomo H, Takahashi T, Ito H, Saito T (1999) Biosynthesis and characterization of poly(3-hydroxybutyrate-*co*-3-hydroxyvalerate) produced by *Burkholderia cepacia* D1. Int J Biol Macromol 24:311–318

Mohamad H, Ridzuan R, Anis M, Hasamudin WWH, Kamarudin H, Ropandi M, Aztimar AA (2005) Research and development of oil palm biomass in wood based industries. Malaysian Palm oil board bulletin/palm oil developments 36/information series

Mokhtari-Hosseini ZB, Vasheghani-Farahani E, Heidarzadeh-Vazifekhoran A, Shojaosadati SA, Karimzadeh R, Darani KK (2009) Statistical media optimization for growth and PHB production from methanol by a methylotrophic bacterium. Bioresour Technol 100:2436–2443

Mori K, Mukoyama S, Zhang Y, Sato H, Ozaki Y, Terauchi H, Noda I, Takahashi I (2008) Crystalline lamellae and surface morphology of biodegradable polyhydroxyalkanoate thin films: thermal behavior and comparison between poly(3-hydroxybutyrate-*co*-3-hydroxyhexanoate) and poly(3-hydroxybutyrate). Macromolecules 41:1713–1719

Mosior M, Newton AC (1995) Mechanism of interaction of protein kinase C with phorbol esters. J Biol Chem 270:25526–25533

Malaysian Palm Oil Board, MPOB (2007a) Annual production of oil palm products: 1975–2007. Available online at http://econ.mpob.gov.my/economy/annual/stat2007/Production3.2.htm. Accessed Oct 2010

Malaysian Palm Oil Board, MPOB (2007b) Average oil yield for oil palm estates: 2007. Available online at http://econ.mpob.gov.my/economy/annual/stat2007/Area1.12.htm. Accessed Oct 2010

Malaysian Palm Oil Board, MPOB (2007c) Overview of the Malaysia oil palm industry: 2007. Available online at http://econ.mpob.gov.my/economy/EID_web.htm. Accessed Oct 2010

Malaysian Palm Oil Board, MPOB (2008). World production of 17 oils & fats: 1999–2008. Available online at http://econ.mpob.gov.my/economy/annual/stat2008/World6.3.pdf. Accessed Oct 2010

Malaysian Palm Oil Board, MPOB (2009). Overview of the Malaysian oil palm industry 2009. Available online at http://econ.mpob.gov.my/economy/Overview_2009.pdf. Accessed Oct 2010

Mujumdar AM, Misar AV (2004) Anti-inflammatory activity of *Jatropha curcas* roots in mice and rats. J Ethnopharmacol 90:11–15

Murphy DJ (2004) Overview of applications of plant biotechnology. In: Christou P, Klee H (eds) Handbook of plant biotechnology. Wiley, New York, pp 35–48

Murphy DJ (2007) Future prospects for oil palm in the 21st century: biological and related challenges. Eur J Lipid Sci Technol 109:296–306

Murray RGE, Doetsch RN, Robinow CF (1994) Determinative and cytological light microscopy. In: Gerhardt P, Murray RGE, Wood WA, Krieg NR (eds) Manual of methods for general bacteriology, vol 10. American Society for Microbiology, Washington, pp 21–41

Nagapudi K, Brinkman WT, Thomas BS, Park JO, Srinivasarao M, Wright E, Conticello VP, Chaikof EL (2005) Viscoelastic and mechanical behavior of recombinant protein elastomers. Biomaterials 26:4695–4706

Nair LS, Laurencin CT (2007) Biodegradable polymers as biomaterials. Prog Polym Sci 32:762–798

Nath LK, Dutta SK (1992) Wound healing response of the proteolytic enzyme curcain. Ind J Pharmacol 24:114–115

Naylor LA, Wood JC (1999) Process for the microbiological production of pha-polymers. US Patent 5871980

Ndreu A, Nikkola L, Ylikauppila H, Ashammakhi N, Hasirci V (2008) Electrospun biodegradable nanofibrous mats for tissue engineering. Nanomedicine 3:45–60

Nelson T, Kaufman E, Kline E, Sokoloff L (1981) The extraneural distribution of gammahydroxybutyrate. J Neurochem 37:1345–1388

Ng KS, Ooi WY, Goh LK, Shenbagarathai R, Sudesh K (2010) Evaluation of Jatropha oil to produce poly(3-hydroxybutyrate) by *Cupriavidus necator* H16. Polym Degrad Stab 95:1365–1369

Nielsen L (2007) Polyhydroxyalkanoate production in sugarcane - recognizing temporospatial complexity. J Biotechnol 131:S28–S29

Nishida H, Tokiwa Y (1993) Distribution of poly(β-hydroxybutyrate) and poly(ε-caprolactone) aerobic degrading microorganisms in different environments. J Environ Polym Degrad 1:227–233

Noda I, Green PR, Satkowski MM, Schechtman LA (2005) Preparation and properties of a novel class of polyhydroxyalkanoate copolymers. Biomacromolecules 6:580–586

Nonato RV, Mantelatto PE, Rossell CEV (2001) Integrated production of biodegradable plastic, sugar and ethanol. Appl Microbiol Biotechnol 57:1–5

O'Leary W (1962) The fatty acids of bacteria. Bacteriol Rev 26:421–447

Oeding V, Schlegel HG (1973) Beta-ketothiolase from *Hydrogenomonas eutropha* H16 and its significance in the regulation of poly-beta-hydroxybutyrate metabolism. Biochem J 134:239–248

Ogino H, Katou Y, Akagi R, Mimitsuka T, Hiroshima S, Gemba Y, Doukyu N, Yasuda M, Ishimi K, Ishikawa H (2007) Cloning and expression of gene, and activation of an organic solvent-stable lipase from *Pseudomonas aeruginosa* LST-03. Extremophiles 11:809–817

Ohko Y, Utsumi Y, Cen N, Tatsuma T, Kobayakawa K, Satoh Y et al (2001) Self-sterilizing and self-cleaning of silicone catheters coated with TiO₂ photocatalyst thin films: a preclinical work. J Biomed Mater Res Part B: Appl Biomater 58:97–101

Ojumu TV, Yu J, Solomon BO (2004) Production of polyhydroxyalkanoates, a bacterial biodegradable polymer. Afr J Biotechnol 3:18–24

Omar S, Rayes A, Eqaab A, Voß I, Steinbüchel A (2001) Optimization of cell growth and poly (3-hydroxybutyrate) accumulation on date syrup by a *Bacillus megaterium* strain. Biotechnol Lett 23:1119–1123

Openshaw K (2000) A review of *Jatropha curcas*: an oil plant of unfulfilled promise. Biomass Bioenerg 19:1–15

Orts WJ, Nobes GAR, Kawada J, Nguyen S, Yu G, Ravenelle F (2008) Poly(hydroxyalkanoates): biorefinery polymers with a whole range of applications. The work of Robert H Marchessault. Can J Chem 86:628–640

Ostle AG, Holt JG (1982) Nile blue A as a fluorescent stain for poly-beta-hydroxybutyrate. Appl Environ Microbiol 44:238–241

OW (2007) Oil world. Online, Available from World Wide Web http://www.oilworld.biz/. Accessed 1 Oct 2011

Pan P, Inoue Y (2009) Polymorphism and isomorphism in biodegradable polyesters. Prog Polym Sci 34:605–640

Pandian SRK, Deepak V, Kalishwaralal K, Rameshkumar N, Jeyaraj M, Gurunathan S (2010) Optimization and fed-batch production of PHB utilizing dairy waste and sea water as nutrient sources by *Bacillus megaterium* SRKP-3. Bioresour Technol 101:705–711

Pantazaki AA, Papaneophytou CP, Pritsa AG, Liakopoulou-Kyriakides M, Kyriakidis DA (2009) Production of polyhydroxyalkanoates from whey by *Thermus thermophilus* HB8. Process Biochem 44:847–853

Papaneophytou CP, Velali EE, Pantazaki AA (2011) Purification and characterization of an extracellular medium-chain length polyhydroxyalkanoate depolymerase from Thermus thermophilus HB8. Polym Degrad Stab 96:670–678

Park K, Ju YM, Son JS, Ahn KD, Han DK (2007) Surface modification of biodegradable electrospun nanofiber scaffolds and their interaction with fibroblasts. J Biomater Sci Polym Ed 18:369–382

Park S, Yang T, Kang H, Lee S, Lee E, Kim T (2008a) New polyhydroxyalkanoate synthase mutant, useful for preparing lactate polymer or lactate copolymer. WO2008062999-A1; KR2008047279-A

Park SH, Lee SH, Lee EJ, Kang HO, Kim TW, Yang TH, Lee SY (2008b) Copolymer containing 3-hydroxyalkanoate unit and lactate unit, and its manufacturing method. WO/2008/062996

Park SJ, Yang TH, Kang HO, Lee SH, Lee EJ, Kim TW, Lee SY (2008c) Mutants of PHA synthase from Pseudomonas sp.6-19 and method for preparing lactate homopolymer or copolymer using the same. US 2010/0050298 A1

Park WH, Jeong L, Yoo DI, Hudson S (2004) Effect of chitosan on morphology and conformation of electrospun silk fibroin nanofibers. Polymers 45:7151–7157

Patnaik PR (2006) Dispersion optimization to enhance PHB production in fed-batch cultures of Ralstonia eutropha. Bioresour Technol 97:1994–2001

Patra SN, Bhattacharyya D, Ray S, Easteal AJ (2009) Electrospun poly(lactic acid) based conducting nanofibrous networks. IOP Conference Series: Mater Sci Eng doi:10.1088/1757-899X/4/1/012020

Peh KSH, Sodhi NS, de Jong J, Sekercioglu CH, Yap CAM, Lim SLH (2006) Conservation value of degraded habitats for forest birds in southern Peninsular Malaysia. Divers Distrib 12:572–581

Philip S, Keshavarz T, Roy I (2007) Polyhydroxyalkanoates: biodegradable polymers with a range of applications. J Chem Technol Biotechnol 82:233–247

Pierce L, Schroth MN (1994) Detection of pseudomonas colonies that accumulate poly-beta-hydroxybutyrate on Nile blue medium. Plant Dis 78:683–685

Pijuan M, Casàs C, Baeza JA (2009) Polyhydroxyalkanoate synthesis using different carbon sources by two enhanced biological phosphorus removal microbial communities. Process Biochem 44:97–105

Poirier Y (2002) Polyhydroxyalknoate synthesis in plants as a tool for biotechnology and basic studies of lipid metabolism. Prog Lipid Res 41:131–155

Poirier Y, Dennis D, Klomparens K, Nawrath C, Sommerville C (1992a) Perspectives on the production of polyhydroxyalkanoates in plants. FEMS Microbiol Rev 103:237–246

Poirier Y, Dennis DE, Klomparens K, Somerville C (1992b) Polyhydroxybutyrate, a biodegradable thermoplastic, produced in transgenic plants. Science 256:520–523

Pötter M, Madkour MH, Mayer F, Steinbüchel A (2002) Regulation of phasin expression and polyhydroxyalkanoate (PHA) granule formation in Ralstonia eutropha H16. Microbiology 148:2413–2426

Pötter M, Steinbüchel A (2005) Poly(3-hydroxybutyrate) granule-associated proteins: impacts on poly(3-hydroxybutyrate) synthesis and degradation. Biomacromolecules 6:552–560

Pötter M, Steinbüchel A (2006) Biogenesis and structure of polyhydroxyalkanoate granules. In: Shively M (ed) Inclusions in prokaryotes, vol 1. Microbiology Monographs. Springer, Berlin, pp 109–136

Poullin L, Skalski V, Wainberg MA (1987) Effect of phorbol ester on growth of tumors induced by Rous sarcoma virus and on pp 60src kinase activity in these tumors. Cancer Res 47:3637–3642

Povolo S, Toffano P, Basaglia M, Casella S (2010) Polyhydroxyalkanoates production by engineered Cupriavidus necator from waste material containing lactose. Bioresour Technol 101:7902–7907

Pozo C, MartÃ-nez-Toledo MV, Rodelas B, GonzÃ¡lez-LÃ³pez J (2002) Effects of culture conditions on the production of polyhydroxyalkanoates by Azotobacter chroococcum H23 in media containing a high concentration of alpechÃ-n (wastewater from olive oil mills) as primary carbon source. J Biotechnol 97:125–131

Prado AGS, Costa LL (2009) Photocatalytic decouloration of malachite green dye by application of TiO2 nanotubes. J Hazard Mater 169:297–301

Pramanik K (2003) Properties and use of jatropha curcas oil and diesel fuel blends in compression ignition engine. Renew Energy 28:239–248

Qin L-F, Gao X, Liu Q, Wu Q, Chen G-Q (2007) Biosynthesis of polyhydroxyalkanoate copolyesters by Aeromonas hydrophila mutant expressing a low-substrate-specificity PHA synthase PhaC2Ps. Biochem Eng J 37:144–150

Qing TW, Yang ZM, Cai SX, Xu SR, Wu ZZ (1999) Interaction of cell adhesion to materials in tissue engineering. Chin Reparative Reconstruct Surg 13:31–37

Qiu YZ, Han J, Chen GQ (2006) Metabolic engineering of Aeromonas hydrophila for the enhanced production of poly(3-hydroxybutyrate-co-3-hydroxyhexanoate). Appl Microbiol Biotechnol 69:537–542

Qu XH, Wu Q, Liang J, Qu X, Wang SG, Chen GQ (2005) Enhanced vascular-related cellular affinity on surface modified copolyesters of 3-hydroxybutyrate and 3-hydroxyhexanoate (PHBHHx). Biomaterials 26:6991–7001

Qu XH, Wu Q, Liang J, Zou B, Chen GQ (2006) Effect of 3-hydroxyhexanoate content in poly(3-hydroxybutyrate-*co*-3-hydroxyhexanoate) on in vitro growth and differentiation of smooth muscle cells. Biomaterials 27:2944–2950

Quyen DT, Dao TT, Thanh Nguyen SL (2007) A novel esterase from *Ralstonia* sp. M1: Gene cloning, sequencing, high-level expression and characterization. Protein Express Purif 51:133–140

Quyen DT, Le TTG, Nguyen TT, Oh T-K, lee J-K (2005) High-level heterologous expression and properties of a novel lipase from *Ralstonia* sp. M1. Protein Express Purif 39:97–106

Rahuman AA, Gopalakrishnan G, Venkatesan P, Geetha K (2008) Larvicidal activity of some Euphorbiaceae plant extracts against *Aedes aegypti* and *Culex quinquefasciatus* (Diptera: Culicidae). Parasitol Res 102:867–873

Rao U, Sridhar R, Sehgal PK (2010) Biosynthesis and biocompatibility of poly(3-hydroxybutyrate-co-4-hydroxybutyrate) produced by *Cupriavidus necator* from spent palm oil. Biochem Eng J 49:13–20

Ravindranath N, Reddy MR, Mahender G, Ramu R, Kumar KR, Das B (2004) Deoxypreussomerins from *Jatropha curcas*: are they also plant metabolites? Phytochemistry 65:2387–2390

Reddy CSK, Ghai R, Rashmi Kalia VC (2003) Polyhydroxyalkanoates: an overview. Bioresour Technol 87:137–146

Reddy NR, Pierson MD (1994) Reduction in antinutritional and toxic components in plant foods by fermentation. Food Res Int 27:281–290

Rehm BHA (2007) Biogenesis of microbial polyhydroxyalkanoate granules: a platform technology for the production of tailor-made bioparticles. Curr Issues Mol Biol 9:41–62

Reijnders L, Huijbregts MAJ (2008) Palm oil and the emission of carbon-based greenhouse gases. J Clean Prod 16:477–482

Ren J, Liu W, Zhu J, Gu S (2008) Preparation and characterization of electrospun, biodegradable membranes. J Appl Polym Sci 109:3390–3397

Ren Q, Ruth K, Thony-Meyer L, Zinn M (2007) Process engineering for production of chiral hydroxycarboxylic acids from bacterial polyhydroxyalkanoates. Macromol Rapid Commun 28:2131–2136

Renner G, Pongratz K, Braunegg G (1996) Production of poly(3-hydroxybutyrate-co-4-hydroxybutyrate) by Comamonas testosteronii A3. Food Technol Biotechnol 34:91–95

Riis V, Mai W (1988) Gas chromatographic determination of poly-β-hydroxybutyric acid in microbial biomass after hydrochloric acid propanolysis. J Chromatogr 445:285–289

Rincón J, Camarillo R, Rodríguez L, Ancillo V (2010) Fractionation of used frying oil by supercritical CO_2 and cosolvents. Ind Eng Chem Res 49:2410–2418

RMRDC (2004) Raw materials research and development council. Report on survey of selected agricultural raw materials in Nigeria

Robinson J (2004) Squaring the circle? Some thoughts on the idea of sustainable development. Ecol Econ 48:369–384

Rodgers M, Wu G (2010) Production of polyhydroxybutyrate by activated sludge performing enhanced biological phosphorus removal. Bioresour Technol 101:1049–1053

Rosenau F, Jaeger K-E (2000) Bacterial lipases from *Pseudomonas*: Regulation of gene expression and mechanisms of secretion. Biochimie 82:1023–1032

Ruan L, Wang Y, Wai L, Hoi-fu Y (2007) Microcalorimetric research on recombinant *Escherichia coli* with high production of polyhydroxyalkanoates (PHAs). J Therm Anal Calorim 89:953–956

Rug M, Sporer F, Wink M, Liu SY, Henning R, Ruppel A (1997) Molluscicidal properties of *J. curcas* against vector snails of the human parasites *Schistosoma mansoni* and *S. japonicum*. In: Gübits GM, Mittelbach M, Trabi M (eds) Biofuels and industrial products from *J. curcas*. Dbv-Verlag Univ. Graz, pp 227–232

Ryu HW, Cho KS, Kim BS, Chang YK, Chang HN, Shim HJ (1999) Mass production of poly(3-hydroxybutyrate) by fed-batch cultures of Ralstonia eutropha with nitrogen and phosphate limitation. J Microbiol Biotechnol 9:751–756

Sabra W, Abou-Zeid DM (2008) Improving feeding strategies for maximizing polyhdroxybutyrate yield by *Bacillus megaterium*. Res J Microbiol 3:308–318

Sacristan J, Reinecke H, Mijangos C (2000) Surface modification of PVC films in solvent-nonsolvent mixtures. Polymer 41:5577–5582

Saito M, Inoue Y, Yoshie N (2001) Cocrystallization and phase segregation of blends ofpoly (3-hydroxybutyrate)and poly(3-hydroxybutyrate-*co*-3-hydroxyvalerate). Polymer 42:5573–5580

Saito Y, Nakamura S, Hiramitsu M, Doi Y (1996) Microbial synthesis and properties of poly(3-hydr1oxybutyrate-co-4-hydroxybutyrate). Polym Int 39:169–174

Salimon J, Abdullah R (2008) Physicochemical properties of Malaysian *Jatropha curcas* seed oil. Sains Malaysiana 37:379–382

Salmond CV, Kroll RG, Booth IR (1984) The effect of food preservatives on pH homeostasis in *Escherichia coli*. J Gen Microbiol 130:2845–2850

Sambanthamurthi R, Sundram K, A. TY (2000a) Chemistry and biochemistry of palm oil. Prog Lipid Res 39:507–558

Sambanthamurthi R, Sundram K, Tan Y (2000) Biochemistry of Palm Oil. Prog Lipid Res 39:507–558

Sang B-I, Hori K, Tanji Y, Unno H (2001) A kinetic analysis of the fungal degradation process of poly(3-hydroxybutyrate-co-3-hydroxyvalerate) in soil. Biochem Eng J 9:175–184

Sathesh PC, Murugesan AG (2010) Effective utilization and management of coir industrial waste for the production of poly-β-hydroxybutyrate (PHB) using the bacterium *Azotobacter Beijerinickii*. Int J Environ Res 4:519–524

Satoh H, Mino T, Matsuo T (1992) Uptake of organic substrates and accumulation of polyhydroxyalkanoates linked with glycolysis of intracellular carbohydrates under anaerobic conditions in the biological excess phosphate removal processes. Water Sci Technol 26:933–942

Savenkova L, Gercberga Z, Bibers I, Kalnin M (2000) Effect of 3-hydroxy valerate content on some physical and mechanical properties of polyhydroxyalkanoates produced by *Azotobacter chroococcum*. Process Biochem 36:445–450

Scandola M, Ceccorulli G, Doi Y (1990) Viscoelastic relaxations and thermal properties of bacterial poly(3-hydroxybutyrate-*co*-3-hydroxyvalerate) and poly(3-hydroxybutyrate-*co*-4-hydroxybutyrate). Int J Biol Macromol 12:112–-117

Scharlemann JPW, Laurance WF (2008) How green are biofuels? Science 319:43–44

Senior PJ, Dawes EA (1973) The regulation of poly-beta-hydroxybutyrate metabolism in Azotobacter beirjerinckii. Biochem J 134:225–238

Shah AA, Hasan F, Hameed A (2010) Degradation of poly(3-hydroxybutyrate-co-3-hydroxyvalerate) by a newly isolated Actinomadura sp. AF-555, from soil. Int Biodeterior Biodegrad 64:281–285

Shah S, Sharma A, Gupta MN (2005) Extraction of oil from *Jatropha curcas* L. seed kernels by combination of ultrasonication and aqueous enzymatic oil extraction. Bioresour Technol 96:121–123

Shahhosseini S (2004) Simulation and optimisation of PHB production in fed-batch culture of *Ralstonia eutropha*. Process Biochem 39:963–969

Shahidi F (ed) (2005) Bailey's industrial oil and fat products, vol 6. 6th edn. Wiley, Hoboken

Shang L, Jiang M, Yun Z, Yan HQ, Chang HN (2008) Mass production of medium-chain-length poly(3-hydroxyalkanoates) from hydrolyzed corn oil by fed-batch culture of Pseudomonas putida. World J Microbiol Biotechnol 24:2783–2787

Shang L, Yim SC, Park HG, Chang HN (2004) Sequential feeding of glucose and valerate in a fed-batch culture of *Ralstonia eutropha* for production of poly(hydroxybutyrate-*co*-hydroxyvalerate) with high 3-hydroxyvalerate fraction. Biotechnol Prog 20:140–144

Shangguan YY, Wang YW, Wu Q, Chen GQ (2006) The mechanical properties and in vitro biodegradation and biocompatibility of UV-treated poly(3-hydroxybutyrate-*co*-3-hydroxyhexanoate). Biomaterials 27:2349–2357

Shen L, Worrell E, Patel M (2010) Present and future development in plastics from biomass. Biofuels, Bioprod Biorefining 4:25–40

Shen X-W, Yang Y, Jian J, Wu Q, Chen G-Q (2009) Production and characterization of homopolymer poly(3-hydroxyvalerate) (PHV) accumulated by wild type and recombinant *Aeromonas hydrophila* strain 4AK4. Bioresour Technol 100:4296–4299

Shenoy SL, Bates WD, Frisch HL, Wnek G (2005a) Role of chain entanglements on fiber formation during electrospinning of polymer solutions: good solvent, non-specific polymer–polymer interaction limit. Polymer 46:3372–3384

Shenoy SL, Bates WD, Wnek G (2005b) Correlations between electrospinnability and physical gelation. Polymer 46:8990–9004

Sheu D-S, Lee C-Y (2004) Altering the substrate specificity of polyhydroxyalkanoate synthase 1 derived from *Pseudomonas putida* GPo1 by localized semirandom mutagenesis. J Bacteriol 186:4177–4184

Shi B, Wu W, Wen J, Shi Q, Wu S (2010) Cloning and expression of a lipase gene from *Bacillus subtilis* FS1403 in *Escherichia coli*. Ann Microbiol 60:399–404

Shi H, Shiraishi M, Shimizu K (1997) Metabolic flux analysis for biosynthesis of poly (β-hydroxybutyric acid) in *Alcaligenes eutrophus* from various carbon sources. J Ferment Bioeng 84:579–587

Shin M, Yoshimoto H, Vacanti JP (2004) In vivo bone tissue engineering using mesenchymal stem cells on a novel electrospun nanofibrous scaffold. Tissue Eng 10:33–41

Shinomiya M, Iwata T, Doi Y (1998) The adsorption of substrate-binding domain of PHB depolymerases to the surface of poly(3-hydroxybutyric acid). Int J Biol Macromol 22:129–135

Sim YC, Sudesh K (Unpublished). Annual report 2009 (2009) http://www.simedarby.com/downloads/pdfs/SDB/Annual_Report/Sime_Darby_AR2009.pdf. Accessed online, Accessed Oct 1 2010

Snell KD, Peoples OP (2009) PHA bioplastic: A value-added coproduct for biomass biorefineries. Biofuels, Bioprod Biorefining 3:456–467

Sodhi NS, Koh LP, Brook BW, Ng PKL (2004) Southeast Asian biodiversity: an impending disaster. Trends Ecol Evol 19:654–660

Solaiman D, Ashby R, Foglia T, Marmer W (2006) Conversion of agricultural feedstock and coproducts into poly(hydroxyalkanoates). Appl Microbiol Biotechnol 71:783–789

Solaiman DK, Ashby RD, Foglia TA (2002) Physiological characterization and genetic engineering of Pseudomonas corrugata for medium-chain-length polyhydroxyalkanoates synthesis from triacylglycerols. Curr Microbiol 44:189–195

Song C, Zhao L, Ono S, Shimasaki C, Inoue M (2001) Production of poly(3-hydroxybutyrate-*co*-3-hydroxyvalerate) from cottonseed oil and valeric acid in batch culture of *Ralstonia* sp. strain JC-64. Appl Biochem Biotechnol 94:169–178

Song JH, Jeon CO, Choi MH, Yoon SC, Park W (2008) Polyhydroxyalkanoate (PHA) production using waste vegetable oil by *Pseudomonas* sp strain DR2. J Microbiol Biotechnol 18:1408–1415

Song S, Hein S, Steinbuchel A (1999) Production of poly(4-hydroxybutyric acid) by fed-batch cultures of recombinant strains of *Escherichia coli*. Biotechnol Lett 21:193–197

Sozer N, Kokini JL (2009) Nanotechnology and its applications in the food sector. Trends Biotechnol 27:82–89

Sparks J, Scholz C (2008) Synthesis and characterization of a cationic poly(β-hydroxybutyrate). Biomacromol 9:2091–2096

Spiekermann P, Rehm BHA, Kalscheuer R, Baumeister D, Steinbüchel A (1999) A sensitive, viable-colony staining method using Nile red for direct screening of bacteria that accumulate polyhydroxyalkanoic acids and other lipid storage compounds. Arch Microbiol 171:73–80

Sridewi N, Bhubalan K, Sudesh K (2006) Degradation of commercially important polyhydroxyalkanoates in tropical mangrove ecosystem. Polym Degrad Stab 91:2931–2940

Steinbüchel A, Lütke-Eversloh T (2003) Metabolic engineering and pathway construction for biotechnological production of relevant polyhydroxyalkanoates in microorganisms. Biochem Eng J 16:81–96

Steinbüchel A (2005) Non-biodegradable biopolymers from renewable resources: perspectives and impacts. Curr Opin Biotechnol 16:607–613

Steinbüchel A, Füchtenbusch B (1998) Bacterial and other biological systems for polyester production. Trends Biotechnol 16:419–427

Steinbüchel A, Valentin HE (1995) Diversity of bacterial polyhydroxyalkanoic acids. FEMS Microbiol Lett 128:219–228

Steinbüchel A, Schlegel HG (1991) Physiology and molecular genetics of poly(beta-hydroxyalkanoic acid) synthesis in *Alcaligenes eutrophus*. Mol Microbiol 5:535–542

Stubbe J, Tian J (2003) Polyhydroxyalkanoate (PHA) homeostasis: the role of the PHA synthase. Nat Prod Rep 20:445–457

Subbiah T, Bhat GS, Tock RW, Parameswaran S, Ramkumar SS (2005) Electrospinning of nanofibers. J Appl Polym Sci 96:557–569

Subramanian A, Lin HY, Vu D, Larsen G (2004) Synthesis and evaluation of scaffolds prepared from chitosan fibers for potential use in cartilage tissue engineering. Biomed Sci Instrum 40:117–122

Sudesh K (2004) Microbial polyhydroxyalkanoates (PHAs): an emerging biomaterial for tissue engineering and therapeutic applications. Med J Malays 59 Suppl B:55–56

Sudesh K, Abe H (2010) Practical guide to microbial polyhydroxyalkanoates. Smithers Rapra Technology, UK

Sudesh K, Abe H, Doi Y (2000) Synthesis, structure and properties of polyhydroxyalkanoates: biological polyesters. Prog Polym Sci 25:1503–1555

Sudesh K, Bhubalan K, Chuah J-A, Kek Y-K, Kamilah H, Sridewi N, Lee Y-F (2011) Synthesis of polyhydroxyalkanoate from palm oil and some new applications. Appl Microbiol Biotechnol 89:1373–1386

Sudesh K, Doi Y (2005) Polyhydroxyalkanoates. Handbook of Biodegradable Polymers

Sudesh K, Iwata T (2008) Sustainability of biobased and biodegradable plastics.Clean 36:433–442

Sudesh K, Loo CY, Goh LK, Iwata T, Maeda M (2007) The oil-absorbing property of polyhydroxyalkanoate films and its practical application: a refreshing new outlook for an old degrading material. Macromol Biosci 7:1199–1205

Sudesh, K, Taguchi K, Doi Y (2001) Can cyanobacteria be a potential PHA producer? RIKEN Rev. No. 42:75–76

Sun T, Norton D, McKean RJ, Haycock JW, Ryan AJ, MacNeil S (2007) Development of a 3D cell culture system for investigating cell interactions with electrospun fibers. Biotechnol Bioeng 97:1318–1328

Sunada K, Watanabe T, Hashimoto K (2003) Studies on photokilling of bacteria on TiO_2 thin film. J Photochem Photobiol A: Chem 156:227–233

Suriyamongkol P, Weselake R, Narine S, Moloney M, Shah S (2007) Biotechnological approaches for the production of polyhydroxyalkanoates in microorganisms and plants – A review. Biotechnol Adv 25:148–175

Suwantong O, Waleetorncheepsawat S, Sanchavanakit N, Pavasant P, Cheepsunthorn P, Bunaprasert T, Supaphol P (2007) In vitro biocompatibility of electrospun poly(3-hydroxybutyrate) and poly(3-hydroxybutyrate-co-3-hydroxyvalerate) fiber mats. Int J Biol Macromol 40:217–223

Suzuki T, Sugawa E, Yano T, Nomoto T, Imamura T, Honma T, Kenmoku T (2004) Polyhydroxyalkanoate polyester having vinyl phenyl structure in the side chain and its production method. Tokyo, US Patent 6803444

Taguchi K, Taguchi S, Sudesh K, Maehara A, Tsuge T, Doi Y (2004) Metabolic pathways and engineering of PHA biosynthesis. In: Steinbüchel A, Doi Y (eds) Biotechnology of biopolymers: from synthesis to patents. Wiley-VCH, pp 217–247

Taguchi S, Doi Y (2004) Evolution of polyhydroxyalkanoate (PHA) production system by "enzyme evolution": Successful case studies of directed evolution. Macromol Biosci 4:145–156

Taguchi S, Tsuge T (2008) Natural polyester-related proteins: structure, function, evolution and engineering. Protein Engineering Handbook. Wiley-VCH, pp 877–914

Taguchi S, Yamada M, Matsumoto K, Tajima K, Satoh Y, Munekata M, Ohno K, Kohda K, Shimamura T, Kambe H, Obata S (2008) A microbial factory for lactate-based polyesters using a lactate-polymerizing enzyme. Proc Nat Acad Sci US Am 105:17323–17327

Takagi Y, Hashii M, Maehara A, Yamane T (1999) Biosynthesis of polyhydroxyalkanoate with a thiophenoxy side group obtained from *Pseudomonas putida*. Macromolecules 32:8315–8318

Takagi Y, Yasuda R, Maehara A, Yamane T (2004) Microbial synthesis and characterization of polyhydroxyalkanoates with fluorinated phenoxy side groups from *Pseudomonas putida*. Eur Polym J 40:1551–1557

Takeoka GR, Full GH, Dao LT (1997) Effect of heating on the characteristics and chemical composition of selected frying oils and fats. J Agric Food Chem 45:3244–3249

Tamer IM, Murray MY, Chisti Y (1998) Optimization of poly(β-hydroxybutyric acid) recovery from *Alcaligenes latus*: combined mechanical and chemical treatments. Bioproc Biosyst Eng 19:459–468

Tan IKP, Sudesh Kumar K, Theanmalar M, Gan SN, Gordon Iii B (1997) Saponified palm kernel oil and its major free fatty acids as carbon substrates for the production of polyhydroxyalkanoates in *Pseudomonas putida* PGA1. Appl Microbiol Biotechnol 47:207–211

Tan KB, Obendorf SK (2007) Fabrication and evaluation of electrospun nanofibrous antimicrobial nylon 6 membranes J Membr Sci 305:287–298

Tan KT, Lee KT, Mohamed AR, Bhatia S (2009) Palm oil: addressing issues and towards sustainable development. Renew Sustain Energy Rev 13:420–427

Tanadchangsaeng N, Kitagawa A, Yamamoto T, Abe H, Tsuge T (2009) Identification, biosynthesis and characterization of polyhydroxyalkanoate copolymer consisting of 3-hydroxybutyrate and 3-hydroxy-4-methylvalerate. Biomacromolecules 10:2866–2874

Tang HY, Ishii D, Mahara A, Murakami S, Yamaoka T, Sudesh K, Samian R, Fujita M, Maeda M, Iwata T (2008) Scaffolds from electrospun polyhydroxyalkanoate copolymers: Fabrication, characterization, bioabsorption and tissue response. Biomaterials 29:1307–1317

Taniguchi I, Kagotani K, Kimura Y (2003) Microbial production of poly(hydroxyalkanoate)s from waste edible oils. Green Chem 5:545–548

Taylor GI (1964) Disintegration of water drops in an electric field. Proc R Soc Lond A 280:383–397

Thompson RC, Swan SH, Moore CJ, Vom Saal FS (2009) Our plastic age. Philos Trans R Soc B: Biol Sci 364:1973–1976

Tian J, Sinskey AJ, Stubbe J (2005) Kinetic studies of polyhydroxybutyrate granule formation in *Wautersia eutropha* H16 by transmission electron microscopy. J Bacteriol 187:3814–3824

Tokiwa Y, Calabia BP (2004) Degradation of microbial polyesters. Biotechnol Lett 26:1181–1189

Tokiwa Y, Ugwu CU (2007) Biotechnological production of (R)-3-hydroxybutyric acid monomer. J Biotechnol 132:264–272

Tong H-W, Wang M (2007) Electrospinning of aligned biodegradable polymer fibers and composite fibers for tissue engineering applications. J Nanosci Nanotechnol 7:3834–3840

Tsuge T (2002) Metabolic improvements and use of inexpensive carbon sources in microbial production of polyhydroxyalkanoates. J Biosci Bioeng 94:579–584

Tsuge T, Saito Y, Kikkawa Y, Hiraishi T, Doi Y (2004) Biosynthesis and compositional regulation of poly[(3-hydroxybutyrate)-*co*-(3-hydroxyhexanoate)] in recombinant *Ralstonia eutropha* expressing mutated polyhydroxyalkanoate synthase genes. Macromol Biosci 4:238–242

Tsuge T, Taguchi K, Taguchi S, Doi Y (2003) Molecular characterization and properties of (R)-specific enoyl-CoA hydratases from *Pseudomonas aeruginosa*: metabolic tools for synthesis of polyhydroxyalkanoates via fatty acid β-oxidation. Int J Biol Macromol 31:195–205

Tsuge T, Tanaka K, Ishizaki A (2001) Development of a novel method for feeding a mixture of -lactic acid and acetic acid in fed-batch culture of Ralstonia eutropha for poly–3-hydroxybutyrate production. J Biosci Bioeng 91:545–550

Tsuge T, Tanaka K, Shimoda M, Ishizaki A (1999) Optimization of -lactic acid feeding for the production of poly–3-hydroxybutyric acid by Alcaligenes eutrophus in fed-batch culture. J Biosci Bioeng 88:404–409

Tsuge T, Yano K, Imazu S-i, Numata K, Kikkawa Y, Abe H, Taguchi S, Doi Y (2005) Biosynthesis of polyhydroxyalkanoate (PHA) copolymer from fructose using wild-type and laboratory-evolved PHA synthases. Macromol Biosci 5:112–117

Turesin F, Gursel I, Hasirci V (2001) Biodegradable polyhydroxyalkanoate implants for osteomyelitis therapy: in vitro antibiotic release. J Biomater Sci Polym Ed 12:195–207

USDA (2005) Oilseed: World Markets and Trade. (Circular Series):FOP 9-05

Valappil SP, Peiris D, Langley GJ, Herniman JM, Boccaccini AR, Bucke C, Roy I (2007) Polyhydroxyalkanoate (PHA) biosynthesis from structurally unrelated carbon sources by a newly characterized *Bacillus* spp. J Biotechnol 127:475–487

Valentin H, Schonebaum A, Steinbuchel A (1992) Identification of 4-hydroxyvaleric acid as a constituent in biosynthetic polyhydroxyalkanoic acids from bacteria. Appl Microbiol Biotechnol 36:507–514

Valentin HE, Dennis D (1997) Production of poly(3-hydroxybutyrate-co- 4-hydroxybutyrate) in recombinant *Escherichia coli* grown on glucose. J Biotechnol 58:33–38

Valentin HE, Lee EY, Choi CY, Steinbuchel A (1994) Identification of 4-hydroxyhexanoic acid as a new constituent of biosynthetic polyhydroxyalkanoic acids from bacteria. Appl Microbiol Biotechnol 40:710–716

Valentin HE, Schonebaum A, Steinbuchel A (1996) Identification of 5-hydroxyhexanoic acid, 4-hydroxyheptanoic acid and 4-hydroxyoctanoic acid as new constituents of bacterial polyhydroxyalkanoic acids. Appl Microbiol Biotechnol 46:261–267

Valentin HE, Steinbuchel A (1993) Application of enzymatically synthesized short-chain-length hydroxyl fatty acid coenzyme A thioesters for assay of polyhydroxyalkanoic acid synthases. Appl Microbiol Biotechnol 40:699–709

Valentin HE, Zwingmann G, Schonebaum A, Steinbuchel A (1995) Metabolic pathway for biosynthesis of poly(3-hydroxybutyrate-co-4-hydroxybutyrate) from 4-hydroxybutyrate by Alcaligenes eutrophus. Eur J Biochem 227:43–60

Venugopal J, Ma LL, Yong T, Ramakrishna S (2005) In vitro study of smooth muscle cells on polycaprolactone and collagen nanofibrous matrices. Cell Biol Int 29:861–867

Verlinden RAJ, Hill DJ, Kenward MA, Williams CD, Radecka I (2007) Bacterial synthesis of biodegradable polyhydroxyalkanoates. J Appl Microbiol 102:1437–1449

Vidal-Mas J, Resina P, Haba E, Comas J, Manresa A, Vives-Rego J (2001) Rapid flow cytometry–Nile red assessment of PHA cellular content and heterogeneity in cultures of Pseudomonas aeruginosa 47T2 (NCIB 40044) grown in waste frying oil. Antonie Van Leeuwenhoek 80:57–63

Viswanathan MB, Ramesh N, Ahilan A, Lakshmanaperumalsamy P (2004) Phytochemical constituents and antimicrobial activity from the stems of *Jatropha maheshwarii*. Med Chem Res 13:361–368

Volova TG, Boyandin AN, Vasiliev AD, Karpov VA, Prudnikova SV, Mishukova OV, Boyarskikh UA, Filipenko ML, Rudnev VP, Bá Xuân B, Vi, t Dung V, Gitelson II (2010) Biodegradation of polyhydroxyalkanoates (PHAs) in tropical coastal waters and identification of PHA-degrading bacteria. Polym Degrad Stab 95:2350–2359

Volova TG, Kalacheva GS (2005) The synthesis of hydroxybutyrate and hydroxyvalerate copolymers by the bacterium *Ralstonia eutropha*. Microbiology 74:54–59

Wallen LL, Rohwedder WK (1974) Poly-β-hydroxyalkanoate from activated sludge. Environ Sci Technol 8:576–579

Waltermann M, Luftmann H, Baumeister D, Kalscheuer R, Steinbüchel A (2000) *Rhodococcus opacus* strain PD630 as a new source of high-valuesing-cell oil? Isolation and characterization of triacylglycerols and other storage lipids. Microbiology 146:1143–1149

Wang C, Hsu CH, Hwang IH (2008) Scaling laws and internal structure for characterizing electrospun poly[(R)-3-hydroxybutyrate] fibers. Polymer 49:4188–4195

Wang F, Lee SY (1997) Poly(3-hydroxybutyrate) production with high productivity and high polymer content by a fed-batch culture of *Alcaligenes latus* under nitrogen limitation. Appl Environ Microbiol 63:3703–3706

Wang J, Yue Z-B, Sheng G-P, Yu H-Q (2010) Kinetic analysis on the production of polyhydroxy-alkanoates from volatile fatty acids by *Cupriavidus necator* with a consideration of substrate inhibition, cell growth, maintenance, and product formation. Biochem Eng J 49:422–428

Wang Y, Yamada S, Asakawa N, Yamane T, Yoshie N, Inoue Y (2001) Comonomer compositional distribution and thermal and morphological characteristics of bacterial poly(3-hydroxybu-tyrate-*co*-3-hydroxyvalerate)s with high 3-hydroxyvalerate content. Biomacromolecules 2:1315–1323

Wang YW, Mo W, Yao H, Wu Q, Chen J, Chen GQ (2004) Biodegradation studies of poly(3-hydroxybutyrate-co-3-hydroxyhexanoate). Polym Degrad Stab 85:815–821

Wang YW, Wu Q, Chen GQ (2003) Reduced mouse fibroblast cell growth by increased hydro-philicity of microbial polyhydroxyalkanoates via hyaluronan coating. Biomaterials 24:4621–4629

Wang YW, Wu Q, Chen J, Chen GQ (2005a) Evaluation of three-dimensional scaffolds made of blends of hydroxyapatite and poly(3-hydroxybutyrate-co-3-hydroxyhexanoate) for bone reconstruction. Biomaterials 26:899–904

Wang YW, Yang F, Wu Q, Cheng YC, Yu PHF, Chen J, Chen GQ (2005b) Effect of composition of poly(3-hydroxybutyrate-co-3-hydroxyhexanoate) on growth of fibroblast and osteoblast. Biomaterials 26:755–761

Weiss TJ (1983) Food oils and their uses. In: Morton JF (ed) Economic botany. AVI Publishing Company, Westport, Connecticut, p 432

Wender PA, Kee J-M, Warrington JM (2008) Practical synthesis of prostratin, DPP, and their ana-logs, adjuvant leads against latent HIV. Science 320:649–652

Werner C, Freier T (2006) Biopolyesters in tissue engineering applications. In: Polymers for regenerative medicine, vol 203. Adv Polym Sci. Springer, Berlin, pp 1–61

Wilhelm S, Gdynia A, Tielen P, Rosenau F, Jaeger K-E (2007) The autotransporter esterase EstA of *Pseudomonas aeruginosa* is required for rhamnolipid production, cell motility, and biofilm formation. J Bacteriol 189:6695–6703

Wilhelm S, Tommassen J, Jaeger K-E (1999) A novel lipolytic enzyme located in the outer mem-brane of *Pseudomonas aeruginosa*. J Bacteriol 181:6977–6986

Williams SF, Martin DP (2002) Applications of PHAs in medicine and pharmacy. In: Steinbüchel A (ed) Series of biopolymers in 10 volumes, vol 4. Wiley/VCY, Weinheim, pp 91–121

Williams SF, Martin DP, Horowitz DM, Peoples OP (1999) PHA applications: addressing the price performance issue I. tissue engineering. Int J Biol Macromol 25:111–121

Wink M, Koschmieder C, Sauerwein M, Sporer F (1997) Phorbol esters of *J. curcas* −biological activities and potential applications. In: Gübits GM, Mittelbach M, Trabi M (eds) Biofuels and industrial production from *J. curcas*. Dbv-Verlag Univ. Graz, pp 160–166

WWF (2005) Agricultural and environment: Palm oil; habitat conservation. Online, Accessed 1 Oct 2011 (Available from World Wide Web), http://www.panda.org/about_wwwf/what_ we_do/policy/agriculture_environment/commodities/palm_oil/environmental_impacts/ habitat_conversion/index.cfm

Xie WP, Chen G-Q (2008) Production and characterization of terpolyester poly(3-hydroxybu-tyrate-*co*-4-hydroxybutyrate-*co*-3-hydroxyhexanoate) by recombinant *Aeromonas hydrophila* 4AK4 harboring genes *phaPCJ*. Biochem Eng J 38(3):384–389

Xu C, Yang F, Wang S, Ramakrishna S (2004) In vitro study of human vascular endothelial cell function on materials with various surface roughness. J Biomed Mater Res A 71:154–161

Xu Y, Wang RH, Koutinas AA, Webb C (2010) Microbial biodegradable plastic production from a wheat-based biorefining strategy. Process Biochem 45:153–163

Yamanaka K, Kimura Y, Aoki T, Kudo T (2010) Effect of ethylene glycol on the end group struc-ture of poly(3-hydroxybutyrate). Polym Degrad Stabil 95:1284–1291

Yang F, Murugan R, Wang S, Ramakrisha S (2005) Electrospinning of nano/micro scale poly (L-lactic acid) aligned fibers and their potential in neural tissue engineering. Biomaterials 26:2603–2610

Yang F, Xu CY, Kotaki M, Wang S, Ramakrishna S (2004) Characterization of neural stem cells on electrospun poly(L-lactic acid) nanofibrous scaffold. J Biomater Sci Polym Ed 15:1483–1497

Yang XS, Zhao K, Chen GQ (2002) Effect of surface treatment on the biocompatibility of microbial polyhydroxyalkanoates. Biomaterials 23:1391–1397

Yang Y, Zhang C, Xu Y, Wang H, Li X, Wang C (2010) Electrospun Er:TiO$_2$ nanofibrous films as efficient photocatalysts under solar simulated light. Mater Lett 64:147–150

Yano T, Nomoto T, Kozaki S, Imamura T, Honma T (2009) Polyhydroxyalkanoate-containing magnetic structure, and manufacturing method and use thereof. US Patent 7527809

Yano T, Sugawa E, Imamura T, Honma T, Kenmoku T (2003) Polyhydroxyalkanoate containing unit with thienyl structure in the side chain, process for its production, charge control agent, toner binder and toner which contain this polyhydroxyalkanoate, and image-forming method and image-forming apparatus which make use of the toner. US Patent 6777153

Yates RA, Caldwell JD (1993) Regeneration of oils used for deep frying: a comparison of active filter aids. J Am Oil Chem Soc 70:507–511

Ye M, Li C, Francis G, Makkar H (2009) Current situation and prospects of *Jatropha curcas* as a multipurpose tree in China. Agroforest Syst 76:487–497

Yew SP, Tang HY, Sudesh K (2006) Photocatalytic activity and biodegradation of polyhydroxybutyrate films containing titanium dioxide. Polym Degrad Stab 91:1800–1807

York GM, Stubbe J, Sinskey AJ (2001) New insight into the role of PhaP phasing of *Ralstonia eutropha* in promoting synthesis of polyhydroxybutyrate. J Bacteriol 183:2394–2397

Yoshie N, Menju H, Sato H, Inoue Y (1995) Complex composition distribution of poly(3-hydroxybutyrate-*co*-3-hydroxyvalerate). Macromolecules 28:6516–6521

Yoshie N, Saito M, Inoue Y (2001) Structural transition of lamella crystals in a isomorphous copolymer, poly(3-hydroxybutyrate-*co*-3-hydroxyvalerate). Macromolecules 34:8953–8960

Yu B-Y, Chen P-Y, Sun Y-M, Lee Y-T, Young T-H (2008) The behaviors of human mesenchymal stem cells on the poly(3-hydroxybutyrate-*co*-3-hydroxyhexanoate) (PHBHHx) membranes. Desalination 234:204–211

Yu F, Dong T, Zhu B, Tajima K, Yazawa K, Inoue Y (2007) Mechanical properties of comonomer-compositionally fractionated poly[(3-hydroxybutyrate)-*co*-(3-mercaptopropionate)] with low 3-mercaptopropionate unit content. Macromol Biosci 7:810–819

Yu J (2001) Production of PHA from starchy wastewater via organic acids. J Biotechnol 86:105–112

Yu J, Si Y, Wong WKR (2002) Kinetics modeling of inhibition and utilization of mixed volatile fatty acids in the formation of polyhydroxyalkanoates by *Ralstonia eutropha*. Process Biochem 37:731–738

Yuan W, Jia Y, Tian J, Snell K, Müh U, Sinskey A et al (2001) Class I and III polyhydroxyalkanoate synthases from *Ralstonia eutropha* and *Allochromatium vinosum*: characterization and substrate specificity studies. Arc Biochem Biophys 394:87–98

Yusoff S (2006) Renewable energy from palm oil—Innovation on effective utilization of waste. J Cleaner Prod 14:87–93

Žagar E, Kržan A, Adamus G, Kowalczuk M (2006) Sequence distribution in microbial poly(3-hydroxybutyrate-*co*-3-hydroxyvalerate) co-polyesters determined by NMR and MS. Biomacromolecules 7:2210–2216

Zakaria MR, Ariffin H, Mohd Johar NA, Abd-Aziz S, Nishida H, Shirai Y, Hassan MA (2010) Biosynthesis and characterization of poly(3-hydroxybutyrate-co-3-hydroxyvalerate) copolymer from wild-type Comamonas sp. EB172. Polym Degrad Stab 95:1382–1386

Zakaria MR, Tabatabaei M, Ghazali FM, Abd-Aziz S, Shirai Y, Hassan MA (2009) Polyhydroxyalkanoate production from anaerobically treated palm oil mill effluent by new bacterial strain *Comamonas* sp. EB172. World J Microbiol Biotechnol 26:767–774

Zarkoob S, Eby RK, Reneker DH, Hudson SD, Ertley D, Adams WW (2004) Structure and morphology of electrospun silk nanofibers. Polymer 45:3973–3977

Zhang H-F, Ma L, Wang Z-H, Chen G-Q (2009a) Biosynthesis and characterization of 3-hydroxyalkanoate terpolyesters with adjustable properties by *Aeromonas hydrophila*. Biotechnol Bioeng 104(3):582–589

Zhang H, Obias V, Gonyer K, Dennis D (1994) Production of polyhydroxyalkanoates in sucrose-utilizing recombinant *Escherichia coli* and *Klebsiella* strains. Appl Environ Microbiol 60:1198–1205

Zhang X, Xu S, Han G (2009b) Fabrication and photocatalytic activity of TiO_2 nanofiber membrane. Mater Lett 63:1761–1763

Zhang YZ, Huang JVZM, Lim CT, Ramakrishna S (2005) Characterization of the surface biocompatibility of the electrospun PCL-collagen nanofibers using fibroblasts. Biomacromolecules 6:2583–2589

Zhao J, Geuskens G (1999) Surface modification of polymers VI. Thermal and radiochemical grafting of acrylamide on polyrthylene and polystyrene. Eur Polym J 35:2115–2123

Zhao K, Deng Y, Chen GQ (2003) Effects of surface morphology on the biocompatibility of polyhydroxyalkanoates. Biochem Eng J 16:115–123

Zhao W, Chen G-Q (2007) Production and characterization of terpolyester poly(3-hydroxybutyrate-*co*-3-hydroxyvalerate-*co*-3-hydroxyhexanoate) by recombinant *Aeromonas hydrophila* 4AK4 harboring genes *phaAB*. Process Biochem 42:1342–1347

Zheng J, He A, Li J, Xu J, Han CC (2006) Studies on the controlled morphology and wettability of polystyrene surfaces by electrospinning or electrospraying. Polymer 47:7095–7102

Zheng Z, Bei FF, Tian HL, Chen GQ (2005) Effects of crystallization of polyhydroxyalkanoate blend on surface physicochemical properties and interactions with rabbit articular cartilage chondrocytes. Biomaterials 26:3537–3548

Zhong SP, Teo WE, Zhu X, Beuerman R, Ramakrishna S, Yung LYL (2007) Development of a novel collagen-GAG nanofibrous scaffold via electrospinning. Mater Sci Eng., C 27:262–266

Zhou T, Lu X, Wang J, Wong FS, Li Y (2008) Rapid decolorization and mineralization of simulated textile wastewater in a heterogeneous Fenton like system with/without external energy. J Hazard Mater 165:193–199

Zinn M, Witholt B, Egli T (2001) Occurrence, synthesis and medical application of bacterial polyhydroxyalkanoate. Adv Drug Deliv Rev 53:5–21

Zong X, Bien H, Chung C-Y, Yin L, Fang D, Hsiao BS, Chu B, Entcheva E (2005) Electrospun fine-textured scaffolds for heart tissue constructs. Biomaterials 26:5330–5338